Alfred Barnard Basset

An elementary treatise on hydrodynamics and sound

Alfred Barnard Basset

An elementary treatise on hydrodynamics and sound

ISBN/EAN: 9783337278458

Printed in Europe, USA, Canada, Australia, Japan

Cover: Foto ©berggeist007 / pixelio.de

More available books at **www.hansebooks.com**

AN ELEMENTARY TREATISE

ON

HYDRODYNAMICS AND SOUND

BY

A. B. BASSET, M.A., F.R.S.

TRINITY COLLEGE, CAMBRIDGE.

CAMBRIDGE:
DEIGHTON, BELL AND CO.
LONDON: GEORGE BELL AND SONS.
1890

PREFACE.

THE treatise on Hydrodynamics, which I published in 1888, was intended for the use of those who are acquainted with the higher branches of mathematics, and its aim was to present to the reader as comprehensive an account of the whole subject as was possible. But although a somewhat formidable battery of mathematical artilleiy is indispensable to those who desire to possess an exhaustive knowledge of any branch of mathematical physics, yet there are a variety of interesting and important investigations, not only in Hydrodynamics, but also in Electricity and other physical subjects, which are well within the reach of every one, who possesses a knowledge of the elements of the Differential and Integral Calculus and the fundamental principles of Dynamics. I have accordingly, in the present work, abstained from introducing any of the more advanced methods of analysis, such as Spherical Harmonics, Elliptic Functions and the like; and, as regards the dynamical portion of the subject, I have endeavoured to solve the various problems which present themselves, by the aid of the Principles of Energy and Momentum, and have avoided the use of Lagrange's equations. There are a few problems, such as the helicoidal steady motion and stability of a solid of revolution moving in an infinite liquid, which cannot be conveniently treated without having recourse to moving axes; but as the theory of moving axes is not an altogether easy branch of Dynamics, I have as far as possible abstained from introducing them, and the reader who is unacquainted with the use of moving axes is recommended to omit those sections in which they are employed.

The present work is principally designed for those who are reading for Part I. of the Mathematical Tripos, under the new regulations, and for other examinations in which an elementary knowledge of Hydrodynamics and Sound is required; but I also trust that it will be of service, not only to those who have neither the time nor the inclination to become conversant with the intricacies of the higher mathematics, but that it will also prepare the way for the acquisition of more elaborate knowledge, on the part of those who have an opportunity of devoting their attention to the more recondite portions of these subjects.

The first part, which relates to Hydrodynamics, has been taken with certain alterations and additions from my larger treatise, and the analytical treatment has been simplified as much as possible. I have thought it advisable to devote a chapter to the discussion of the motion of spheres and circular cylinders, in which the equations of motion are obtained by the direct method of calculating the resultant pressure exerted by the liquid upon the solid; inasmuch as this method is far more elementary, and does not necessitate the use of Green's Theorem, nor involve any further knowledge of Dynamics on the part of the reader, than the ordinary equations of motion of a rigid body. The methods of this chapter can also be employed to solve the analogous problem of determining the electrostatic potential of cylindrical and spherical conductors and accumulators, and the distribution of electricity upon such surfaces. The theory of the motion of a solid body and the surrounding liquid, regarded as a single dynamical system, is explained in Chapter III., and the motion of an elliptic cylinder in an infinite liquid, and the motion of a circular cylinder in a liquid bounded by a rigid plane, are discussed at length.

The Chapter on Waves and on Rectilinear Vortex Motion comprises the principal problems which admit of treatment by elementary methods, and I have also included an investigation due to Lord Rayleigh, respecting one of the simpler cases of the instability of fluid motion.

In the second part, which deals with the Theory of Sound, I have to acknowledge the great assistance which I have received from Lord Rayleigh's classical treatise. This part contains the solution of the simpler problems respecting the vibrations of strings, membranes, bars and gases; and I have also added a few pages on the statical problem of the flexion of bars. A few sections are also devoted to the Thermodynamics of perfect gases, principally for the sake of supplementing Maxwell's treatise on Heat, by giving a proof of some results which require the use of the Differential Calculus.

I have to express my best thanks to Professor Greenhill for having read the proof sheets, and for having made many valuable suggestions during the progress of the work.

CONTENTS.

PART I.

HYDRODYNAMICS.

CHAPTER I.

ON THE EQUATIONS OF MOTION OF A PERFECT FLUID.

ART.		PAGE
1.	Introduction	1
2.	Definition of a fluid	1
3.	Kinematical theorems. Lagrangian and flux methods	2
4.	Velocity and acceleration. The Lagrangian method	2
5.	do. The flux method	3
6.	The equation of continuity	4
7.	The velocity potential	5
8.	Molecular rotation	6
9–10.	Lines of flow and stream lines	6
11.	Earnshaw's and Stokes' current function	7
12.	The bounding surface	8
13.	Dynamical theorems. Pressure at every point of a fluid is equal in all directions	8
14.	The equations of motion	9
15–16.	Another proof of the equations of motion	10
17.	Pressure is a function of the density	13
18.	Equations satisfied by the components of molecular rotation	13
19.	Stokes' proof that a velocity potential always exists, if it exists at any particular instant	14
20.	Physical distinction between rotational and irrotational motion	15
21.	Integration of the equations of motion when a velocity potential exists	16
22.	Steady motion. Bernoulli's theorem	16
23.	Impulsive motion	17
24.	Flow and circulation	18

B. H. b

CONTENTS.

ART.		PAGE
25.	Cyclic and acyclic irrotational motion. Circulation is independent of the time	19
26.	Velocity potential due to a source	20
27.	do. due to a doublet	21
28.	do. due to a source in two dimensions	21
29.	do. due to a doublet in two dimensions	22
30.	Theory of images	22
31.	Image of a source in a plane	23
32.	Image of a doublet in a sphere, whose axis passes through the centre of the sphere	23
33.	Motion of a liquid surrounding a sphere, which is suddenly annihilated	25
34.	Torricelli's theorem	26
35.	Application of the hypothesis of parallel sections, and the motion of liquid flowing out of a vessel	27
36.	The vena contracta	28
37.	Giffard's injector	29
Examples	31

CHAPTER II.

MOTION OF CYLINDERS AND SPHERES IN AN INFINITE LIQUID.

38. Statement of problems to be solved 35
39. Boundary conditions for a cylinder moving in a liquid 36
40. Velocity potential and current function due to the motion of a circular cylinder in an infinite liquid 38
41. Motion of a circular cylinder under the action of gravity . . . 38
42. Motion of a cylinder in a liquid, which is bounded by a concentric cylindrical envelop 40
43. Current function due to the motion of a cylinder, whose cross section is a lemniscate of Bernoulli 41
44. Motion of a liquid contained within an equilateral prism . . . 42
45. do. do. an elliptic cylinder 43
46. Conjugate functions 43
47. Current function due to the motion of an elliptic cylinder . . . 44
48. Failure of solution when the elliptic cylinder degenerates into a lamina. Discontinuous motion 45
49. Motion of a sphere under the action of gravity 47
50. Motion may become unstable owing to the existence of a hollow . . 49
51. Effect of viscosity; and definition of the coefficient of viscosity . . 50
52. Resistance experienced by a ship in moving through water . . . 52
53. Motion of a spherical pendulum, which is surrounded by liquid . . 52
54. Motion of a spherical pendulum, when the liquid is contained within a rigid spherical envelop 53
 Examples 55

CONTENTS. vii

CHAPTER III.

MOTION OF A SINGLE SOLID IN AN INFINITE LIQUID.

ART.   PAGE
55. Fundamental dynamical principles 59
56. Green's theorem 60
57-61. Applications of Green's theorem 61
62. Conditions which the velocity potential must satisfy 63
63. Kinetic energy of liquid is a homogeneous quadratic function of the velocities of the moving solid 64
64. Values of the components of momentum 66
65. Short proof of expressions for the kinetic energy and momentum . . 68
66. Motion of a sphere 68
67-69. Motion of an elliptic cylinder under the action of no forces . . 69
70. Motion of an elliptic cylinder under the action of gravity . . . 74
71. Helicoidal steady motion of a solid of revolution 75
72. Conditions of stability. Application to gunnery 77
73-75. Motion of a circular cylinder parallel to a plane 79
Examples 82

CHAPTER IV.

WAVES.

76. Kinematics of wave motion 85
77. Progressive waves and stationary waves 87
78. Conditions of the problem of wave motion 88
79-80. Waves in a liquid under the action of gravity 89
81-83. Waves at the surface of separation of two liquids 91
84-85. Stable and unstable motion 93
88. Long waves in shallow water 96
89. Analytical theory of long waves 97
90. Stationary waves in flowing water 98
91. Theory of group velocity 99
92. Capillarity 100
93. Capillary waves—conditions at the free surface 101
94. Capillary waves under the action of gravity 101
95.   do.   produced by wind 102
Examples 104

CHAPTER V.

RECTILINEAR VORTEX MOTION.

ART.		PAGE
96.	Definition of a vortex	107
97.	Velocity due to a single vortex	107
98.	Velocity potential due to a vortex	109
99.	Conditions which the pressure must satisfy	109
100.	Kirchhoff's elliptic vortex	111
101.	Discussion on the stability of a vortex	112
102.	Motion of two vortices of equal strengths	113
103.	Motion of two vortices of equal and opposite strengths	114
104.	Motion of a vortex in a square corner	114
105.	Motion of a vortex inside a circular cylinder	115
106.	Rankine's free spiral vortex	116
107.	Fundamental properties of vortex motion	117
	Examples	119

PART II.

THEORY OF SOUND.

CHAPTER VI.

INTRODUCTION.

108.	Noises and musical notes	125
109.	Connection between the characteristics of a note and the geometrical constants of a wave	126
110.	Velocity of propagation of sound in gases and liquids	126
111.	Intensity	126
112.	Pitch	126
113.	Compound notes and pure tones	127
114.	Timbre	128
115.	Beats	128

CHAPTER VII.

VIBRATIONS OF STRINGS AND MEMBRANES.

116.	Transverse and longitudinal vibrations	130
117.	Equation of motion for transverse vibrations	131
118.	Solution for a string whose ends are fixed	132
119.	Initial conditions	133

CONTENTS.   ix

ART.		PAGE
120.	Motion produced by a given displacement	134
121.	Motion produced by an impulse applied at a point	135
122.	Motion produced by a periodic force	136
123.	Free vibrations gradually die away on account of friction	137
124.	Forced vibrations	138
125.	Normal functions. Kinetic and potential energy	138
126.	Longitudinal vibrations	139
127.	Transverse vibrations of membranes	140
128.	Nodal lines of a square membrane	141
129.	Circular membrane	142
	Examples	143

CHAPTER VIII.

FLEXION OF BARS.

130.	Equations of equilibrium of a bar	145
131.	Value of the bending moment	146
132.	Conditions to be satisfied at the ends	148
133.	The elastica	148
134.	Kirchhoff's kinetic analogue	150
135.	Equations of motion of a bar	151
136.	Equation of motion for the lateral vibrations of a naturally straight bar	151
137.	Conditions at a free end	152
138.	Equation of motion and conditions at a free end, when the rotatory inertia is neglected	152
139.	Period of an infinite bar	152
140.	Lateral vibrations of a bar of given length	153
141.	Period equations	153
142.	Longitudinal vibrations	155
143.	Vibrations of a circular bar	156
	Examples	158

CHAPTER IX.

EQUATIONS OF MOTION OF A PERFECT GAS.

144.	Fundamental equations of the small vibrations of a gas	161
145.	Displacement in a plane wave is perpendicular to the wave front	162
146.	Newton's value of the velocity of sound	163
147.	Thermodynamics of gases	164
148.	The second law of Thermodynamics	166
149.	Specific heats of a gas	167
150.	Specific heats of air	167
151.	Equations of the adiabatic and isothermal lines of a perfect gas	168
152.	Elasticity of a perfect gas	169
153.	Velocity of sound in air	170
154.	Intensity of sound	170

CHAPTER X.

PLANE AND SPHERICAL WAVES.

ART.		PAGE
155.	Motion in a closed vessel	172
156–8.	Motion in a cylindrical pipe	173
159.	Reflection and refraction	174
160.	Change of phase, when reflection is total	177
161.	Spherical waves	178
162.	Symmetrical waves in a spherical envelop	179
163.	Waves in a conical pipe	180
164.	Sources of sound	180
165.	Diametral vibrations	180
166.	Motion of a spherical pendulum surrounded by air	181
167.	Scattering of a sound wave by a small rigid sphere	183
	Examples	185
	Note to § 53	188

PART I.

HYDRODYNAMICS.

CHAPTER I.

ON THE EQUATIONS OF MOTION OF A PERFECT FLUID.

1. THE object of the science of Hydrodynamics, is to investigate the motion of fluids. All fluids with which we are acquainted may be divided into two classes, viz. incompressible fluids or liquids, and compressible fluids or gases. It must however be recollected, that all liquids experience a slight compression, when submitted to a sufficiently large pressure, and therefore in strictness a liquid cannot be regarded as an incompressible fluid; but inasmuch as the compression produced by such pressures as ordinarily occur is very small, liquids may be usually treated as incompressible fluids, without sensible error. The physical interest arising from a study of the motion of gases, is due to the fact that air is the vehicle by means of which sound is transmitted. We shall therefore devote the first part of this volume to the discussion of incompressible fluids or liquids, reserving the discussion of gases for the second part, which deals with the Theory of Sound.

We must now define a fluid.

2. *A fluid may be defined to be an aggregation of molecules, which yield to the slightest effort made to separate them from each other, if it be continued long enough.*

A *perfect* fluid, is one which is incapable of sustaining any tangential stress or action in the nature of a shear; and it will be shown in § 13 that the consequence of this property is, that the pressure at every point of a perfect fluid, is equal in all directions, whether the fluid be at rest or in motion. A perfect fluid is however an entirely ideal substance, since all fluids with which we are acquainted are capable of offering resistance to tangential stresses. This property, which is known as viscosity, gives rise to an action in the nature of friction, by which the kinetic energy is gradually converted into heat.

In the case of gases, water and many other liquids, the effects of viscosity are so small that such fluids may be approximately regarded as perfect fluids. The neglect of viscosity very much simplifies the mathematical treatment of the subject, and in the present treatise, we shall confine our attention to perfect fluids.

Before entering upon the dynamical portion of the subject, it will be convenient to investigate certain kinematical propositions, which are true for all fluids.

Kinematical Theorems.

3. The motion of a fluid may be investigated by two different methods, the first of which is called the Lagrangian method, and the second the Eulerian or flux method, although both are due to Euler.

In the Lagrangian method, we fix our attention upon an element of fluid, and follow its motion throughout its history. The variables in this case are the initial coordinates a, b, c of the particular element upon which we fix our attention, and the time. This method has been successfully employed in the solution of very few problems.

In the Eulerian or flux method, we fix our attention upon a particular point of the space occupied by the fluid, and observe what is going on there. The variables in this case are the coordinates x, y, z of the particular point of space upon which we fix our attention, and the time.

Velocity and Acceleration.

4. In forming expressions for the velocity and acceleration of a fluid, it is necessary to carefully distinguish between the Lagrangian and the flux method.

I. *The Lagrangian method.*

Let u, v, w be the component velocities parallel to *fixed* axes, of an element of fluid whose coordinates are x, y, z and $x + \delta x$, $y + \delta y$, $z + \delta z$ at times t and $t + \delta t$ respectively, then

$$u = dx/dt = \dot{x}, \quad v = \dot{y}, \quad w = \dot{z} \quad \ldots\ldots\ldots\ldots(1),$$

where in forming $\ddot{x}$, $\ddot{y}$, $\ddot{z}$ we must suppose x, y, z to be expressed in terms of the initial coordinates a, b, c and the time.

The expressions for the component accelerations are
$$f_x = \dot{u} = \ddot{x}, \quad f_y = \ddot{y}, \quad f_z = \ddot{z} \quad \ldots\ldots\ldots\ldots\ldots(2),$$
where u, v, w are supposed to be expressed in terms of a, b, c and t.

II. *The Flux Method.* EULER

5. Let δQ be the quantity of fluid which in time δt flows across any small area A, which passes through a fixed point P in the fluid; let ρ be the density of the fluid, q its resultant velocity, and ϵ the angle which the direction of q makes with the normal to A, drawn towards the direction in which the fluid flows. Then
$$\delta Q = \rho q \, A \delta t \cos \epsilon,$$
therefore
$$q = \frac{1}{\rho A \cos \epsilon} \frac{dQ}{dt}.$$

Now $A \cos \epsilon$ is the projection of A upon a plane passing through P perpendicular to the direction of motion of the fluid; hence δQ is the independent of the direction of the area, and is the same for all areas whose projections upon the above-mentioned plane are equal. Hence the velocity is equal to the rate per unit of area divided by the density, at which fluid flows across a plane perpendicular to its direction of motion.

The velocity is therefore a function of the position of P and the time.

In the present treatise the flux method will almost be exclusively employed. We may therefore put $u = F(x, y, z, t)$; whence if $u + \delta u$ be the velocity parallel to x at time $t + \delta t$ of the element of fluid which at time t was situated at the point (x, y, z),
$$\delta u = F(x + u\delta t, y + v\delta t, z + w\delta t, t + \delta t) - F(x, y, z, t).$$
Therefore the acceleration,
$$f_x = \lim \frac{\delta u}{\delta t} = \frac{du}{dt} + u\frac{du}{dx} + v\frac{du}{dy} + w\frac{du}{dz}.$$
Hence if $\partial/\partial t$ denotes the operator
$$d/dt + u\,d/dx + v\,d/dy + w\,d/dz,$$
the component accelerations will be given by the equations
$$f_x = \frac{\partial u}{\partial t}, \quad f_y = \frac{\partial v}{\partial t}, \quad f_z = \frac{\partial w}{\partial t} \quad \ldots\ldots\ldots\ldots\ldots(3).$$

4 EQUATIONS OF MOTION OF A PERFECT FLUID.

The Equation of Continuity.

6. If an imaginary fixed closed surface be described in a fluid, the difference between the amounts of fluid which flow in and flow out during a small interval of time δt, must be equal to the increase in the amount of fluid during the same interval, which the surface contains.

The analytical expression for this fact, is called *the equation of continuity*.

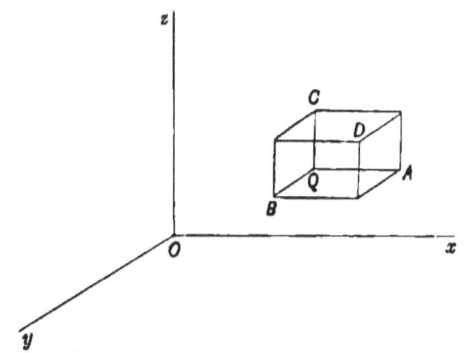

Let Q be any point (x, y, z), and consider an elementary parallelopiped $\delta x \delta y \delta z$.

The amount of fluid which flows in across the face CB in time δt is

$$\rho u \, \delta y \, \delta z \, \delta t.$$

The amount which flows out across the opposite face AD is

$$\rho u \, \delta y \, \delta z \, \delta t + \frac{d}{dx}(\rho u) \, \delta x \, \delta y \, \delta z \, \delta t,$$

whence the gain of fluid due to the fluxes across the faces CB, AD is

$$-\frac{d}{dx}(\rho u) \, \delta x \, \delta y \, \delta z \, \delta t.$$

Treating the other faces in a precisely similar manner, it follows that the total gain is

$$-\left\{\frac{d}{dx}(\rho u) + \frac{d}{dy}(\rho v) + \frac{d}{dz}(\rho w)\right\} \delta x \, \delta y \, \delta z \, \delta t \dots\dots(4).$$

EQUATION OF CONTINUITY. 5

The amount of fluid within the element at time t is $\rho\,\delta x\,\delta y\,\delta z$, and therefore the amount at time $t + \delta t$ is

$$\left(\rho + \frac{d\rho}{dt}\delta t\right)\delta x\,\delta y\,\delta z.$$

The gain is therefore

$$\frac{d\rho}{dt}\delta x\,\delta y\,\delta z\,\delta t.$$

Equating this to (4) we obtain the equation

$$\frac{d\rho}{dt} + \frac{d(\rho u)}{dx} + \frac{d(\rho v)}{dy} + \frac{d(\rho w)}{dz} = 0 \quad\ldots\ldots\ldots\ldots (5).$$

This equation is called the *equation of continuity*.

In the case of a liquid, ρ is constant, and (5) takes the simple form

$$\frac{du}{dx} + \frac{dv}{dy} + \frac{dw}{dz} = 0 \quad\ldots\ldots\ldots\ldots\ldots\ldots\ldots(6).$$

We shall hereafter require the equation of continuity of a liquid referred to polar coordinates. This may be obtained in a similar manner by considering a polar element of volume $r^2\sin\theta\,\delta r\,\delta\theta\,\delta\omega$, and it can be shown that if u, v, w be the velocities in the directions in which r, θ, ω increase, the required equation is

$$\sin\theta\,\frac{d(r^2 u)}{dr} + r\frac{d(v\sin\theta)}{d\theta} + r\frac{dw}{d\omega} = 0 \quad\ldots\ldots\ldots (7).$$

The Velocity Potential.

7. In a large and important class of problems, the quantity $u\,dx + v\,dy + w\,dz$ is a perfect differential of a function of x, y, z which we shall call ϕ; when this is the case, we shall have

$$u\,dx + v\,dy + w\,dz = d\phi,$$

whence $\qquad u = \dfrac{d\phi}{dx}, \quad v = \dfrac{d\phi}{dy}, \quad w = \dfrac{d\phi}{dz} \ldots\ldots\ldots\ldots\ldots (8).$

Substituting these values of u, v, w in (6) we obtain

$$\frac{d^2\phi}{dx^2} + \frac{d^2\phi}{dy^2} + \frac{d^2\phi}{dz^2} = 0 \quad\ldots\ldots\ldots\ldots\ldots\ldots (9),$$

or as it is usually written

$$\nabla^2\phi = 0.$$

This equation is called *Laplace's equation*, from the name of its discoverer; it is a very important equation, which continually occurs in a variety of branches of physics. The operator ∇^2 is called *Laplace's operator*.

We can now obtain the transformation of Laplace's equation when polar coordinates are employed. For in this case

$$udr + vr d\theta + wr \sin\theta d\omega = d\phi,$$

whence $\quad u = \dfrac{d\phi}{dr}, \quad v = \dfrac{1}{r}\dfrac{d\phi}{d\theta}, \quad w = \dfrac{1}{r\sin\theta}\dfrac{d\phi}{d\omega}$ ....(10).

Substituting in (7) we obtain

$$\frac{d}{dr}\left(r^2 \frac{d\phi}{dr}\right) + \frac{1}{\sin\theta}\frac{d}{d\theta}\left(\sin\theta \frac{d\phi}{d\theta}\right) + \frac{1}{\sin^2\theta}\frac{d^2\phi}{d\omega^2} = 0...(11).$$

The equation of continuity and the theory of the velocity potential, may therefore be employed to effect transformations, which it would be very laborious to work out by the usual methods for the change of the independent variables.

8. The existence of a velocity potential involves the conditions that each of the three quantities

$$\frac{dw}{dy} - \frac{dv}{dz}, \quad \frac{du}{dz} - \frac{dw}{dx}, \quad \frac{dv}{dx} - \frac{du}{dy}$$

should be zero; when such is not the case we shall denote these quantities by 2ξ, 2η, 2ζ. The quantities ξ, η, ζ for reasons which will be explained hereafter, are called *components of molecular rotation*, they evidently satisfy the equation

$$\frac{d\xi}{dx} + \frac{d\eta}{dy} + \frac{d\zeta}{dz} = 0 \quad\quad\quad\quad\quad\quad (12).$$

When a velocity potential exists, the motion is called *irrotational*; and when a velocity potential does not exist, the motion is called *rotational or vortex motion*.

Lines of Flow and Stream Lines.

9. DEF. A *line of flow* is a line whose direction coincides with the direction of the resultant velocity of the fluid.

The differential equations of a line of flow are

$$\frac{dx}{u} = \frac{dy}{v} = \frac{dz}{w}.$$

Hence if $\chi_1(x, y, z, t) = a_1$, $\chi_2(x, y, z, t) = a_2$ be any two independent integrals, the equations $\chi_1 = \text{const.}$, $\chi_2 = \text{const.}$, are the equations of two families of surfaces whose intersections determine the lines of flow.

DEF. A *stream line or a line of motion*, is a line whose direction coincides with the direction of the actual paths of the elements of fluid.

The equations of a stream line are determined by the simultaneous differential equations,

$$\dot{x} = u, \quad \dot{y} = v, \quad \dot{z} = w,$$

where x, y, z must be regarded as unknown functions of t. The integration of these equations will determine x, y, z in terms of the initial coordinates and the time.

10. When a velocity potential exists, the equation

$$udx + vdy + wdz = 0$$

is the equation of a family of surfaces, at every point of which the velocity potential has a definite constant value, and which may be called *surfaces of equi-velocity potential*.

If P be any point on the surface, $\phi = $ const., and dn be an element of the normal at P which meets the neighbouring surface $\phi + \delta\phi$ at Q, the velocity at P along PQ, will be equal to $d\phi/dn$; hence $d\phi$ must be positive, and therefore a fluid always flows from places of lower to places of higher velocity potential.

The lines of flow evidently cut the surfaces of equi-velocity potential at right angles.

11. The solution of hydrodynamical problems is much simplified by the use of the velocity potential (whenever one exists), since it enables us to express the velocities in terms of a single function ϕ. But when a velocity potential does not exist, this cannot in general be done, unless the motion either takes place in two dimensions, or is symmetrical with respect to an axis.

In the case of a liquid, if the motion takes place in planes parallel to the plane of xy, the equation of the lines of flow is

$$udy - vdx = 0 \dotfill (13).$$

The equation of continuity is

$$\frac{du}{dx} + \frac{dv}{dy} = 0,$$

which shows that the left-hand side of (13) is a perfect differential $d\psi$, whence

$$u = \frac{d\psi}{dy}, \quad v = -\frac{d\psi}{dx} \dotfill (14).$$

The function ψ is called Earnshaw's current function.

When the motion takes place in planes passing through the axis of z, the equation of the lines of flow may be written

$$\varpi(wd\varpi - udz) = 0 \quad\quad\quad\quad (15),$$

where ϖ, θ, z are cylindrical coordinates.

The equation of continuity[1] is

$$\frac{d(\varpi u)}{d\varpi} + \varpi\frac{dw}{dz} = 0,$$

which shows that the left-hand side of (15) is a perfect differential $d\psi$, whence

$$w = \frac{1}{\varpi}\frac{d\psi}{d\varpi}, \quad u = -\frac{1}{\varpi}\frac{d\psi}{dz} \quad\quad\quad (16).$$

The function ψ is called Stokes' current function.

The Bounding Surface.

12. Besides the equations which must be satisfied within the interior of a fluid, it is necessary that certain other conditions should be satisfied at the boundary, which depend upon the special problem under consideration.

If the fluid is bounded by a surface, whose equation referred to axes fixed in space is $F(x, y, z, t) = 0$, the normal velocity of the fluid at the surface, must be equal to the normal velocity of the surface, hence the sheet of fluid of which the boundary is composed, must always consist of the same elements of fluid. Hence

$$F(x + u\delta t,\ y + v\delta t,\ z + w\delta t,\ t + \delta t) = 0,$$

and therefore

$$\frac{dF}{dt} + u\frac{dF}{dx} + v\frac{dF}{dy} + w\frac{dF}{dz} = 0 \quad\quad\quad (17).$$

If the boundary is fixed, the condition becomes

$$lu + mv + nw = 0 \quad\quad\quad\quad (18),$$

where l, m, n are the direction cosines of the normal to F.

Dynamical Theorems.

13. It has been already stated, that the pressure at every point of a perfect fluid is equal in all directions, whether the fluid be at rest or in motion. It will now be shown that this property is the consequence of such a fluid being incapable of offering resistance to a tangential stress.

---

[1] The equation of continuity in cylindrical coordinates, may be obtained, as in § 6, by considering the fluxes across the sides of an element $\varpi\delta\varpi\delta\theta\delta z$.

EQUATIONS OF MOTION.

Let $ABCD$ be a small tetrahedron of fluid, and let p, p' be the pressures per unit of area upon the faces ABC and BCD.

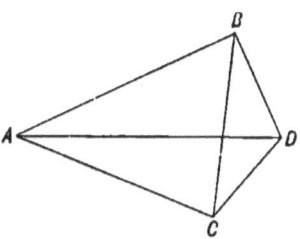

By D'Alembert's Principle, the reversed effective forces and the impressed forces which act upon the volume of fluid, together with the pressures upon its faces, constitute a system in statical equilibrium. The first two vary as the volume, and the last vary as the areas of the faces of the tetrahedron; and therefore if the tetrahedron be made to diminish indefinitely, the former will vanish in comparison with the latter. Hence the tetrahedron will ultimately be in equilibrium under the action of the pressures upon its faces.

Resolve the pressures upon the faces ABC and BCD parallel to AD. Since the projections of the two faces upon a plane perpendicular to AD are equal, the conditions of equilibrium require that $p = p'$, which proves the proposition[1].

The Equations of Motion.

14. The equations of motion of a perfect fluid, may be obtained by two different methods, which we shall proceed to explain.

Let X, Y, Z be the components per unit of mass, of the impressed forces (such as gravity and the like) which act upon the fluid; p the pressure, and ρ the density.

Let Q be any point of the fluid, whose coordinates are x, y, z; and consider an elementary parallelopiped δx, δy, δz (see figure to § 6) whose edges are parallel to the axes.

By D'Alembert's Principle, the reversed effective forces together with the impressed forces, form a system in statical equilibrium. Whence if δm be the mass of an element of fluid contained within this parallelopiped, f_x the component of its acceleration parallel to x, and P_x the resultant parallel to x of the pressure due to the surrounding fluid upon the faces CB, AD; D'Alembert's Principle gives

$$\Sigma (X - f_x) \delta m + P_x = 0 \ \dots\dots\dots\dots\dots(18).$$

[1] This proposition is true even in the case of viscous fluids, provided they are at rest.

10   EQUATIONS OF MOTION OF A PERFECT FLUID.

Now
$$P_x = p\delta y\, \delta z - \left(p + \frac{dp}{dx}\delta x\right)\delta y\, \delta z,$$
whence (18) becomes
$$\Sigma (X - f'_x)\delta m - \frac{dp}{dx}\delta x\, \delta y\, \delta z = 0 \quad \ldots\ldots\ldots\ldots (19).$$

Since the parallelopiped is supposed to be indefinitely small, the first term of this equation becomes in the limit
$$(X - f'_x)\Sigma \delta m = (X - f'_x)\rho\, \delta x\, \delta y\, \delta z,$$
and therefore (19) becomes,
$$\rho X - \frac{dp}{dx} = \rho f'_x \quad \ldots\ldots\ldots\ldots\ldots\ldots (20).$$

In all the applications which will occur, we shall use the flux method, in which case f_x is given by (3); whence (20) may be written
$$X - \frac{1}{\rho}\frac{dp}{dx} = \frac{\partial u}{\partial t}.$$

Resolving parallel to y and z and proceeding in a similar manner, we shall obtain two other equations, which may be deduced by cyclical interchange of the letters x, y, z, and u, v, w respectively; whence the equations of motion are

$$\left.\begin{aligned}X - \frac{1}{\rho}\frac{dp}{dx} &= \frac{du}{dt} + u\frac{du}{dx} + v\frac{du}{dy} + w\frac{du}{dz}\\Y - \frac{1}{\rho}\frac{dp}{dy} &= \frac{dv}{dt} + u\frac{dv}{dx} + v\frac{dv}{dy} + w\frac{dv}{dz}\\Z - \frac{1}{\rho}\frac{dp}{dz} &= \frac{dw}{dt} + u\frac{dw}{dx} + v\frac{dw}{dy} + w\frac{dw}{dz}\end{aligned}\right\}\ldots\ldots(21).$$

15. We shall now obtain the equations of motion by a different method, which will enable us to illustrate some important Dynamical Principles.

It is a well-known Dynamical Principle that—

The rate of change of the component of the linear momentum, parallel to an axis, of any dynamical system, is equal to the component force along that axis.

In order to apply this principle to the motion of a fluid, let $APBQ$ represent the fluid, which at time t, is contained within any imaginary closed surface S, described in the fluid. At the end of an interval δt, the fluid will no longer be contained within

EQUATIONS OF MOTION. 11

S, but will occupy a different position, which is shown by the line $ApBq$ in the figure.

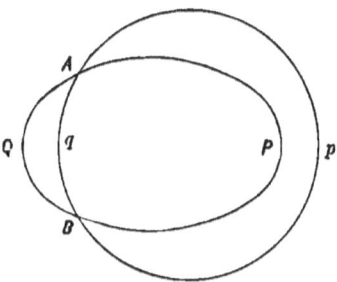

Let M_x, $M_x + \delta M_x$, be the component momenta parallel to x, of the original fluid at times t, $t + \delta t$; X' and P_x the components parallel to x of the impressed forces, and the pressure upon the boundary of the given mass of fluid due to the action of the surrounding fluid.

By the Principle stated above, we obtain immediately the equation

$$\frac{dM_x}{dt} = X' + P_x \quad\quad\quad (22).$$

Let $\delta\mu_1$, $M_x + \delta M'_x$, $\delta\mu_2$ be the component momenta parallel to x, of the fluid which at time $t + \delta t$ occupies the spaces $ApBPA$, $APBQA$ and $AqBQA$; then

$$M_x + \delta M_x = M_x + \delta M'_x + \delta\mu_1 - \delta\mu_2,$$

whence $\quad\quad \delta M_x = \delta M'_x + \delta\mu_1 - \delta\mu_2,$

and therefore (22) may be written

$$\frac{dM'_x}{dt} = \frac{d\mu_2}{dt} - \frac{d\mu_1}{dt} + X' + P_x \quad\quad\quad (23).$$

Now dM'_x/dt is the rate of increase of the component of momentum parallel to x, of the fluid contained within S; and $d\mu_2/dt - d\mu_1/dt$ is the rate at which momentum parallel to x, flows into S. Whence (23) asserts that,—

The rate of increase of the component of momentum parallel to x, of the fluid contained within any given closed surface S, is equal to the rate at which momentum parallel to x flows into x across the boundary of S, together with the rate at which momentum parallel to x, is generated by the component of the impressed force parallel to x, and by the component parallel to x, of the pressure exerted by the surrounding fluid upon the boundary of S.

12 EQUATIONS OF MOTION OF A PERFECT FLUID.

16. In order to apply this proposition, we must calculate the momentum which flows into a surface across its boundary.

Let AB be any element dS of the surface, CA the direction of motion of the fluid, q its resultant velocity, and ϵ the angle which its direction makes with the normal to dS drawn outwards. Through the perimeter of dS, describe a small cylinder, whose curved side contains the lines of flow which pass through the perimeter of dS; then if AM be the projection of dS upon a plane perpendicular to the lines of flow, $BAM = \epsilon$. Hence the total momentum which flows into S across the element dS in time δt is

$$\rho q dS \cos \epsilon . q \delta t = \rho q q' dS \delta t;$$

where $q' = q \cos \epsilon$, is the component velocity perpendicular to dS.

The component in any assigned direction of the momentum which flows into S, is found by multiplying this quantity by the cosine of the angle between this direction and the direction of q; and may therefore be written $\rho q'' q' dS \delta t$, where q'' is the velocity in the given direction.

Applying this to an elementary parallelopiped δx, δy, δz, we see at once that

$$\frac{d\mu_2}{dt} - \frac{d\mu_1}{dt} = -\left\{\frac{d(\rho u^2)}{dx} + \frac{d(\rho uv)}{dy} + \frac{d(\rho uw)}{dz}\right\} \delta x \delta y \delta z.$$

Also
$$\frac{dM'_x}{dt} = \frac{d(\rho u)}{dt} \delta x \delta y \delta z,$$

$$X' = \rho X \delta x \delta y \delta z,$$

whence the equation for motion parallel to x is

$$\rho X - \frac{dp}{dx} = \frac{d(\rho u)}{dt} + \frac{d(\rho u^2)}{dx} + \frac{d(\rho uv)}{dy} + \frac{d(\rho uw)}{dz}.$$

Performing the differentiations on the right-hand side, and taking account of the equation of continuity, this equation reduces to the first of (21).

17. The equations of motion together with the equation of continuity, furnish four relations between the five unknown quantities u, v, w, p, ρ; and are therefore not sufficient to determine the motion.

If however the fluid be a liquid, ρ is constant, and the above-mentioned equations together with the boundary conditions are sufficient to determine the motion; but in the case of a gas another equation is required, which is furnished by means of a relation which exists between p and ρ.

When the motion of the gas is such that the temperature remains constant, we have by Boyle's Law the equation

$$p = k\rho \quad\quad\quad\quad\quad (24),$$

where k is a constant.

But when the motion is such as to cause a sudden compression or dilatation, an increase or decrease of temperature will be produced; and if it is assumed (as is the case with sound waves), that the compression is so sudden that loss or gain of heat by radiation may be neglected, it will be shown in the second part, that the required relation is

$$p = k'\rho^\gamma \quad\quad\quad\quad\quad (25),$$

where γ is the ratio of the specific heat at constant pressure to the specific heat at constant volume. This quantity for all gases has the approximately constant value 1·408.

18. Let us now suppose that the forces arise from a conservative system whose potential is V. Since p is a function of ρ, we may put

$$Q = -\int \frac{dp}{\rho} - V,$$

and the left-hand sides of (21) will be respectively equal to dQ/dx, dQ/dy, dQ/dz. If therefore we eliminate Q by differentiating the second equation with respect to z and the third with respect to y, and introduce the values of ξ, η, ζ from § 8, we shall obtain

$$\frac{\partial \xi}{\partial t} = \xi \frac{du}{dx} + \eta \frac{dv}{dx} + \zeta \frac{dw}{dx} - \xi\theta,$$

where ξ, η, ζ are the components of molecular rotation and $\theta = du/dx + dv/dy + dw/dz$. Eliminating θ by means of the equation of continuity $\partial \rho/\partial t + \rho\theta = 0$, and taking account of the two other equations which may be written down from symmetry, we shall obtain

$$\left.\begin{aligned}\frac{\partial}{\partial t}\left(\frac{\xi}{\rho}\right) &= \frac{\xi}{\rho}\frac{du}{dx}+\frac{\eta}{\rho}\frac{dv}{dx}+\frac{\zeta}{\rho}\frac{dw}{dx}\\ \frac{\partial}{\partial t}\left(\frac{\eta}{\rho}\right) &= \frac{\xi}{\rho}\frac{du}{dy}+\frac{\eta}{\rho}\frac{dv}{dy}+\frac{\zeta}{\rho}\frac{dw}{dy}\\ \frac{\partial}{\partial t}\left(\frac{\zeta}{\rho}\right) &= \frac{\xi}{\rho}\frac{du}{dz}+\frac{\eta}{\rho}\frac{dv}{dz}+\frac{\zeta}{\rho}\frac{dw}{dz}\end{aligned}\right\}\dots\dots\dots(26).$$

These equations may also be written in the form

$$\frac{\partial}{\partial t}\left(\frac{\xi}{\rho}\right)=\frac{\xi}{\rho}\frac{du}{dx}+\frac{\eta}{\rho}\frac{du}{dy}+\frac{\zeta}{\rho}\frac{du}{dz}, \&c. \&c.$$

19. It was stated in § 7, that in many important problems, the motion is such that a velocity potential exists. The condition that such should be the case is, that ξ, η, ζ should each vanish. We shall now prove, that when the fluid is under the action of a conservative system of forces, a velocity potential will always exist whenever it exists at any particular instant.

Let us choose the particular instant at which a velocity potential exists, as the origin of the time; then by hypotheses ξ, η, ζ vanish when $t=0$; also the coefficients of these quantities in (26), will not become infinite at any point of the interior of the fluid; it will therefore be possible to determine a quantity L, which shall be a superior limit to the numerical values of these coefficients. Hence ξ, η, ζ cannot increase faster than if they satisfied the equation

$$\frac{\partial}{\partial t}\left(\frac{\xi}{\rho}\right)=\frac{L}{\rho}(\xi+\eta+\zeta), \&c. \&c.$$

But if $\xi+\eta+\zeta=\Omega\rho$, we obtain by adding the above equations

$$\frac{\partial\Omega}{\partial t}=3L\Omega,$$

whence $\Omega = Ae^{3Lt}$.

Now $\Omega = 0$ when $t=0$, therefore $A=0$; and since Ω is the sum of three quantities each of which is essentially positive, it follows that ξ, η, ζ must always remain zero, if they are so at any particular instant. The above proof is due to Sir G. Stokes[1].

[1] "On the friction of fluids in motion," Section II. *Trans. Camb. Phil. Soc.* vol. VIII.

20. There is, as was first shown by Sir G. Stokes, an important physical distinction in the character of the motion which takes place, according as a velocity potential does or does not exist.

Conceive an indefinitely small spherical element of a fluid in motion to become suddenly solidified, and the fluid about it to be suddenly destroyed. By the instantaneous solidification, velocities will be suddenly generated or destroyed in the different portions of the element, and a set of mutual impulsive forces will be called into action.

Let x, y, z be the coordinates of the centre of inertia G of the element at the instant of solidification, $x + x', y + y', z + z'$ those of any other point P in it; let u, v, w be the velocities of G along the three axes just before solidification, u', v', w' the velocities of P relative to G; also let $\bar{u}, \bar{v}, \bar{w}$ be the velocities of G, u_1, v_1, w_1 the relative velocities of P, and ξ, η, ζ the angular velocities just after solidification. Since all the impulsive forces are internal, we have
$$\bar{u} = u, \quad \bar{v} = v, \quad \bar{w} = w.$$

We have also by the principle of conservation of angular momentum,
$$\Sigma m \{y'(w_1 - w') - z'(v_1 - v')\} = 0, \&c.$$
m denoting an element of the mass of the element considered.

But $u_1 = \eta z' - \zeta y'$, and u' is ultimately equal to
$$\frac{du}{dx}x' + \frac{du}{dy}y' + \frac{du}{dz}z',$$
and similar expressions hold good for the other quantities. Substituting in the above equation, and observing that
$$\Sigma my'z' = \Sigma m'z'x' = \Sigma mx'y' = 0, \text{ and } \Sigma mx'^2 = \Sigma my'^2 = \Sigma mz'^2,$$
we have
$$\xi = \tfrac{1}{2}\left(\frac{dw}{dy} - \frac{dv}{dz}\right), \&c.$$

We see then that an indefinitely small spherical element of the fluid, if suddenly solidified and detached from the rest of the fluid, will begin to move with a motion of translation alone, or a motion of translation combined with one rotation, according as $u\,dx + v\,dy + w\,dz$ is, or is not, an exact differential, and in the latter case the angular velocities will be determined by the equations
$$2\xi = \frac{dw}{dy} - \frac{dv}{dz}, \quad 2\eta = \frac{du}{dz} - \frac{dw}{dx}, \quad 2\zeta = \frac{dv}{dx} - \frac{du}{dy}.$$

On account of the physical meaning of the quantities ξ, η, ζ, they are called the *components of molecular rotation*, and motion which is such that they do not vanish is called *rotational or vortex motion*; when they vanish, the motion is called *irrotational*.

In the foregoing investigation, it has been assumed that the pressure is a function of the density, and also that the fluid is under the action of a conservative system of forces; it therefore follows that vortex motion cannot be produced, and consequently, if once set up, cannot be destroyed by such a system of forces. It can however be shown that the theorem is not true if the pressure is not a function of the density. If therefore by reason of any chemical action, the pressure should cease to be a function of the density during any interval of time however short, vortex motion might be produced, or if in existence might be destroyed.

21. The equations of motion can be integrated whenever a force and a velocity potential exist; for putting

$$Q = -\int \frac{dp}{\rho} - V,$$

and multiplying (21) by dx, dy, dz respectively and adding, we obtain

$$dQ = \frac{\partial u}{\partial t} dx + \frac{\partial v}{\partial t} dy + \frac{\partial w}{\partial t} dz.$$

Now in the present case

$$\frac{\partial u}{\partial t} = \frac{du}{dt} + u \frac{du}{dx} + v \frac{dv}{dx} + w \frac{dw}{dx}$$

$$= \frac{d}{dx} \left(\frac{d\phi}{dt} + \tfrac{1}{2} q^2 \right),$$

where q is the resultant velocity. Integrating, we obtain

$$\int \frac{dp}{\rho} + V + \frac{d\phi}{dt} + \tfrac{1}{2} q^2 = F(t) \quad \ldots\ldots\ldots\ldots\ldots (27),$$

where F is an arbitrary function.

22. When the motion is steady, du/dt, dv/dt and dw/dt are each zero, and in this case the general equations of motion can always be integrated. It will however be necessary to distinguish between irrotational and rotational motion.

BERNOULLI'S THEOREM.

The general equations of motion may be written,

$$\left.\begin{aligned}\frac{\partial u}{\partial t} &= \frac{du}{dt} + \tfrac{1}{2}\frac{dq^2}{dx} - 2v\zeta + 2w\eta = \frac{dQ}{dx} \\ \frac{\partial v}{\partial t} &= \frac{dv}{dt} + \tfrac{1}{2}\frac{dq^2}{dy} - 2w\xi + 2u\zeta = \frac{dQ}{dy} \\ \frac{\partial w}{\partial t} &= \frac{dw}{dt} + \tfrac{1}{2}\frac{dq^2}{dz} - 2u\eta + 2v\xi = \frac{dQ}{dz}\end{aligned}\right\}\ldots\ldots\ldots(28).$$

When the motion is steady and irrotational $\dot u, \dot v, \dot w, \xi, \eta, \zeta$ are each zero; whence multiplying by dx, dy, dz, adding and integrating we obtain

$$Q = \tfrac{1}{2}q^2 - C,$$

or
$$\int\frac{dp}{\rho} + V + \tfrac{1}{2}q^2 = C \ldots\ldots\ldots\ldots\ldots(29).$$

In this case the quantity C is evidently an absolute constant.

When the motion is rotational, let ds be an element of a stream line, then

$$u = q\frac{dx}{ds},\quad v = q\frac{dy}{ds},\quad w = q\frac{dz}{ds}.$$

Multiplying the general equations by u, v, w and adding, we obtain

$$\frac{dQ}{ds} = \tfrac{1}{2}\frac{dq^2}{ds},$$

whence
$$\int\frac{dp}{\rho} + V + \tfrac{1}{2}q^2 = A \ldots\ldots\ldots\ldots\ldots(30).$$

This is Bernoulli's Theorem.

If we use C. G. S. units, the left-hand side of this equation is the energy in ergs per gramme of liquid.

Since we have integrated along a stream line, the quantity A is not an absolute constant, but a function of the parameter of a stream line: in other words if $\psi = $ const., $\chi = $ const. be two surfaces whose intersections determine the stream lines, A is a function of ψ and χ.

Impulsive Motion.

23. The equations which determine the change of motion when a fluid is acted upon by impulsive forces, may be deduced in manner similar to that employed in § 14.

Let u, v, w and u', v', w' be the velocities of the fluid just before and just after the impulse; p the impulsive pressure.

B. H.

18   EQUATIONS OF MOTION OF A PERFECT FLUID.

Since impulsive forces are equal to the change of momentum which they produce, it follows by considering the motion of a small parallelopiped $\delta x \delta y \delta z$, that

$$\rho(u' - u)\,\delta x \delta y \delta z = p\delta y \delta z - \left(p + \frac{dp}{dx}\delta x\right)\delta y \delta z,$$

whence the equations of impulsive motion are

$$\left.\begin{array}{l} \rho(u' - u) = -\dfrac{dp}{dx} \\[4pt] \rho(v' - v) = -\dfrac{dp}{dy} \\[4pt] \rho(w' - w) = -\dfrac{dp}{dz} \end{array}\right\} \quad \ldots\ldots\ldots\ldots(31).$$

Multiplying by dx, dy, dz and adding we obtain

$$-dp/\rho = (u' - u)\,dx + (v' - v)\,dy + (w' - w)\,dz \quad \ldots\ldots(32).$$

In the case of a liquid ρ is constant, whence differentiating with respect to x, y, z, adding and taking account of the equation of continuity, we obtain

$$\nabla^2 p = 0 \quad \ldots\ldots\ldots\ldots\ldots\ldots(33).$$

If the liquid were originally at rest, it is clear that the motion produced by the impulse must be irrotational, whence if ϕ be its velocity potential

$$p = -\rho\phi \quad \ldots\ldots\ldots\ldots\ldots\ldots(34),$$

which is a very important result.

Flow and Circulation.

24. The line integral $\int(u\,dx + v\,dy + w\,dz)$, taken along any curve joining a fixed point A with a variable point P, is called the *flow* from A to P.

If the points A and P coincide, so that the curve along which the integration takes place is a closed curve, this line integral is called the *circulation* round the closed curve.

If the motion of a liquid is irrotational, and ϕ_A, ϕ_P denote the values of the velocity potential at A and P, the flow from A to P is simply $\phi_P - \phi_A$, and is independent of the path from A to P; also the circulation round any closed curve is zero, *provided ϕ be a single-valued function*. Cases however occur in which ϕ is a *many-valued function;* and when this is the case, the value of the circulation will depend upon the position of the closed curve round which the integration is taken, being zero for some curves, whilst for others it has a finite value.

For example, when the motion is in two dimensions, ϕ satisfies the equation
$$\frac{d^2\phi}{dx^2}+\frac{d^2\phi}{dy^2}=0,$$
and it can be verified by trial, that a particular solution of this equation is
$$\phi = m\tan^{-1} y/x.$$
This value of ϕ therefore gives a possible kind of irrotational motion. Let θ be the *least* value of the angle $\tan^{-1} y/x$; then since the equation $\theta = \tan^{-1} y/x$ is satisfied by $\theta + 2n\pi$, where n is any positive or negative integer, it follows that the most general value of ϕ is
$$\phi = m\theta + 2mn\pi,$$
whence ϕ is a many-valued function.

Let a point P start from any position, and describe a closed curve which does not surround the origin. During the passage of P from its original to its final position, the angle θ increases to a certain value, then diminishes, and finally arrives at its original value, and therefore the circulation round such a curve is zero; but if the closed curve surrounds the origin, θ increases from its original value to $2\pi + \theta$, as the point travels round the closed curve, and therefore the circulation round a curve which encloses the origin is $2m\pi$.

Irrotational motion which is characterized by a single-valued velocity potential, is called *acyclic irrotational motion;* whilst motion which is characterized by a many-valued velocity potential, is called *cyclic irrotational motion.*

25. The importance of the distinction between cyclic and acyclic motion will not be fully understood, until we discuss the theory of rectilinear vortex motion; but the results of § 23 will enable us to prove, that cyclic motion cannot be produced or destroyed by impulsive forces.

Integrate (32) round any closed curve, then since p/ρ (or $\int \rho^{-1} dp$ in the case of a gas) is necessarily a single-valued function, it vanishes when integrated round any closed curve, and we obtain
$$\int (u'dx + v'dy + w'dz) = \int (udx + vdy + wdz),$$
which shows that the circulation is unaltered by the impulse.

20   EQUATIONS OF MOTION OF A PERFECT FLUID.

We can also show that cyclic irrotational motion cannot be generated nor destroyed, when the liquid is under the action of forces having a single-valued potential; for if we put

$$H = \int \frac{dp}{\rho} + V + \frac{d\phi}{dt} + \tfrac{1}{2}q^2,$$

the equations of motion are

$$\frac{dH}{dx} = 0, \quad \frac{dH}{dy} = 0, \quad \frac{dH}{dz} = 0.$$

Multiply these equations by dx, dy, dz, add and integrate round a closed curve, and let κ be the circulation; we obtain

$$\int \frac{dp}{\rho} + (V + \tfrac{1}{2}q^2)_2 - (V + \tfrac{1}{2}q_2)_1 + \frac{d\phi}{dt} + \frac{d\kappa}{dt} - \frac{d\phi}{dt} = 0 \ldots(35),$$

where the suffixes refer to the initial and final positions of the moving point. Since $\int \rho^{-1} dp$ and $V + \tfrac{1}{2}q^2$ are single-valued functions, the sum of the first three terms is zero, and (35) reduces to

$$\frac{d\kappa}{dt} = 0,$$

whence $\kappa = \text{const.}$

If therefore κ is zero, or the motion is acyclic, it will remain zero during the subsequent motion.

Sources, Doublets and Images.

26.   When the motion of a liquid is irrotational and symmetrical with respect to a fixed point, which we shall choose as the origin, the value of ϕ at any other point P is a function of the distance alone of P from the origin; and Laplace's equation becomes

$$\frac{d^2\phi}{dr^2} + \frac{2}{r}\frac{d\phi}{dr} = 0.$$

Therefore $\phi = -\dfrac{m}{r},$

and $\dfrac{d\phi}{dr} = \dfrac{m}{r^2}.$

The origin is therefore a singular point, from or to which the stream lines either diverge or converge, according as m is positive or negative. In the former case the singular point is called a *source*, in the latter case a *sink*.

SOURCES AND DOUBLETS. 21

The flux across any closed surface surrounding the origin is,

$$\iint \frac{d\phi}{dr} dS = \iint \frac{m \cos \epsilon}{r^2} dS = m \iint d\Omega$$
$$= 4\pi m,$$

where $d\Omega$ is the solid angle subtended by dS at the origin, and ϵ is the angle which the direction of motion makes with the normal to S drawn outwards.

The constant m is called the strength of the source.

27. A *doublet* is formed by the coalescence of an equal source and sink. To find its velocity potential, let there be a source and sink at S and H respectively, and let O be the middle point of SH, then

$$\phi = -\frac{m}{SP} + \frac{m}{HP}$$
$$= -\frac{mSH \cos SOP}{OP^2}.$$

Now let SH diminish and m increase indefinitely, but so that the product $m . SH$ remains finite and equal to μ, then

$$\phi = -\frac{\mu \cos SOP}{r^2}$$
$$= -\frac{\mu z}{r^3},$$

if the axis of z coincides with OS.

Hence the velocity potential due to a doublet, is equal to the magnetic potential of a small magnet whose axis coincides with the axis of the doublet, and whose negative pole corresponds to the source end of the doublet.

28. When the motion is in two dimensions, and is symmetrical with respect to the axis of z, Laplace's equation becomes

$$\frac{d^2\phi}{dr^2} + \frac{1}{r}\frac{d\phi}{dr} = 0.$$

Therefore $\phi = m \log r,$
$$\frac{d\phi}{dr} = \frac{m}{r},$$

where r is the distance of any point from the axis. This value of ϕ represents a line source of infinite length, whose strength per unit of length is equal to m.

If ψ be the current function,

$$\frac{m}{r} = \frac{1}{r}\frac{d\psi}{d\theta}.$$

Therefore
$$\psi = m\theta$$
$$= m\tan^{-1}\frac{y}{x}.$$

29. The velocity potential due to a doublet in two-dimensional motion is

$$\phi = m\log SP - m\log HP$$
$$= -m\frac{SH}{OP}\cos SOP = -\frac{\mu\cos SOP}{r}$$
$$= -\frac{\mu x}{r^2}.$$

Theory of Images.

30. Let H_1, H_2 be any two hydrodynamical systems situated in an infinite liquid. Since the lines of flow either form closed curves or have their extremities in the singular points or boundaries of the liquid, it will be possible to draw a surface S, which is not cut by any of the lines of flow, and over which there is therefore no flux, such that the two systems H_1, H_2 are completely shut off from one another.

The surface S may be either a closed surface such as an ellipsoid, or an infinite surface such as a paraboloid.

If therefore we remove one of the systems (say H_2) and substitute for it such a surface as S, everything will remain unaltered on the side of S on which H_1 is situated; hence the velocity of the liquid due to the combined effect of H_1 and H_2 will be the same as the velocity due to the system H_1 in a liquid which is bounded by the surface S.

The system H_2 is called the *image* of H_1 with respect to the surface S, and is such that if H_2 were introduced and S removed, there would be no flux across S.

The method of images was invented by Sir William Thomson, and has been developed by Helmholtz, Maxwell and other writers; it affords a powerful method of solving many important physical problems.

31. We shall now give some examples.

Let S, S' be two sources whose strengths are m. Through A the middle point of SS' draw a plane at right angles to SS'. The normal component of the velocity of the liquid at any point P on this plane is

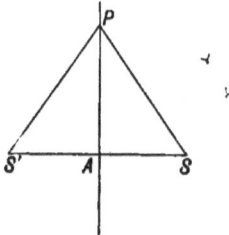

$$-\frac{m}{SP^2}\cos PSA + \frac{m}{S'P^2}\cos PS'A = 0.$$

Hence there is no flux across AP. If therefore Q be any point on the right-hand side of AP, the velocity potential due to a source at S, in a liquid which is bounded by the fixed plane AP, is

$$\phi = -\frac{m}{SQ} - \frac{m}{S'Q}.$$

Hence the image of a source S with respect to a plane is an equal source, situated at a point S' on the other side of the plane, whose distance from it is equal to that of S.

32. The image in a sphere, of a doublet whose axis passes through the centre of the sphere, can also be found by elementary methods.

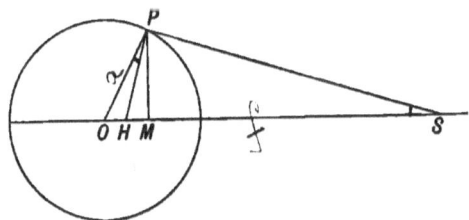

Let S be the doublet, O the centre of the sphere, a its radius, and let $OS = f$.

The velocity potential of a doublet situated at the origin and whose axis coincides with OS, has already been shown to be

$$\phi = -\frac{m\cos\theta}{r^2};$$

whence if R, Θ be the radial and transversal velocities

$$R = \frac{d\phi}{dr} = \frac{2m\cos\theta}{r^3},$$

$$\Theta = \frac{1}{r}\frac{d\phi}{d\theta} = \frac{m\sin\theta}{r^3}.$$

Hence if we have a doublet at S, the component velocity along OP is

$$-\frac{2m}{SP^3}\cos OSP \cos OPS - \frac{m}{SP^3}\sin OSP \sin OPS$$

$$= -\frac{m}{SP^3}\{\cos OSP \cos OPS + \cos (OPS - OSP)\}\ldots\ldots(36).$$

Let us take a point H inside the sphere such that $OH = a^2/f$; then it is known from geometry that the triangles OPH and OSP are similar, and therefore the preceding expression may be written

$$-\frac{m}{SP^3}\{\cos OPH \cos OHP + \cos SPH\}.$$

But the normal velocity due to a doublet of strength m' placed at H is by (36)

$$-\frac{m'}{HP^3}\{\cos OPH \cos OHP + \cos SPH\}$$

and therefore the normal velocity will be zero if

$$\frac{m}{SP^3} + \frac{m'}{HP^3} = 0$$

for all positions of P. But by a well-known theorem,

$$\frac{f}{SP} = \frac{a}{HP},$$

and therefore the condition that the normal velocity should vanish, is that

$$m' = -ma^3/f^3.$$

Whence the image of a doublet of strength m in a liquid bounded by a sphere, is another doublet placed at the inverse point H, whose strength is $-ma^3/f^3$.

The theory of sources, sinks and doublets furnishes a powerful method of solving certain problems relating to the motions of solid bodies in a liquid[1].

We shall conclude this chapter by working out some examples.

---

[1] If a magnetic system be suddenly introduced into the neighbourhood of a conducting spherical shell, it can be shown that the effect of the induced currents at points outside the shell, is *initially* equivalent to a magnetic system inside the shell, which is the hydrodynamical image of the external system; and that the law of decay of the currents, is obtained by supposing the radius of the shell to diminish according to the law $a\epsilon^{-\sigma t/4\pi a}$, where σ is the specific resistance of the shell. Analogous results hold good in the case of a plane current sheet; hence all results concerning hydrodynamical images in spheres and planes, are capable of an electromagnetic interpretation. See C. Niven, *Phil. Trans.* 1881.

EXAMPLES AND APPLICATIONS.   25

33. *A mass of liquid whose external surface is a sphere of radius* a, *and which is subject to a constant pressure* Π, *surrounds a solid sphere of radius* b. *The solid sphere is annihilated, it is required to determine the motion of the liquid.*

It is evident that the only possible motion which can take place, is one in which each element of liquid moves towards the centre, whence the free surfaces will remain spherical. Let R', R be their external and internal radii at any subsequent time, r the distance of any point of the liquid from the centre. The equation of continuity is

$$\frac{d}{dr}(r^2 v) = 0,$$

whence $r^2 v = F(t).$

The equation for the pressure is

$$\frac{1}{\rho}\frac{dp}{dr} = -\frac{dv}{dt} - v\frac{dv}{dr}$$

$$= -\frac{F'(t)}{r^2} - \tfrac{1}{2}\frac{dv^2}{dr}$$

whence $\dfrac{p}{\rho} = A + \dfrac{F'(t)}{r} - \tfrac{1}{2} v^2,$

when $r = R'$, $p = \Pi$, and when $r = R$, $p = 0$, whence if V, V' be the velocities of the internal and external surfaces

$$\frac{\Pi}{\rho} = F'(t)\left(\frac{1}{R'} - \frac{1}{R}\right) - \tfrac{1}{2}(V'^2 - V^2).$$

Since the volume of the liquid is constant,

$$R'^3 - R^3 = a^3 - b^3 = c^3,$$

also $F'(t) = \dfrac{d}{dt}(R^2 V),$

whence

$$\frac{\Pi}{\rho} = V\frac{d}{dR}(R^2 V)\left\{\frac{1}{(R^3+c^3)^{\frac{1}{3}}} - \frac{1}{R}\right\} - \tfrac{1}{2} V^2\left\{\frac{R^4}{(R^3+c^3)^{\frac{4}{3}}} - 1\right\}.$$

Putting $z = R^4 V^2$, multiplying by $2R^2$ and integrating, we obtain

$$\tfrac{2}{3}\frac{\Pi(R^3-b^3)}{\rho R^4} = V^2\left\{\frac{1}{(R^3+c^3)^{\frac{1}{3}}} - \frac{1}{R}\right\},$$

which determines the velocity of the inner surface.

If the liquid had extended to infinity, we must put $c = \infty$, and we obtain

$$\frac{2\Pi}{3\rho}(b^3 - R^3) = R^2\left(\frac{dR}{dt}\right)^2,$$

whence if t be the time of filling up the cavity

$$t = \sqrt{\frac{3\rho}{2\Pi}}\int_0^b \frac{R^{\frac{3}{2}}\,dR}{\sqrt{b^3 - R^3}}.$$

Putting $b^3 x = R^3$, this becomes

$$t = b\sqrt{\frac{\rho}{6\Pi}}\int_0^1 x^{-\frac{1}{6}}(1-x)^{-\frac{1}{2}}\,dx$$

$$= b\sqrt{\frac{\pi\rho}{6\Pi}}\,\frac{\Gamma(\frac{5}{6})}{\Gamma(\frac{4}{3})}.$$

The preceding example may be solved at once by the Principle of Energy.

The kinetic energy of the liquid is

$$2\pi\rho\int_R^{R'} r^2 v^2\,dr = 2\pi\rho V^2 R^4 \int_R^{R'}\frac{dr}{r^2}$$

$$= 2\pi\rho V^2 R^4\left\{\frac{1}{R} - \frac{1}{(R^3 + c^3)^{\frac{1}{3}}}\right\}.$$

The work done by the external pressure is

$$4\pi\Pi\int_R^a r^2\,dr = \tfrac{4}{3}\Pi\pi(a^3 - R'^3)$$

$$= \tfrac{4}{3}\Pi\pi(b^3 - R^3),$$

whence $\quad \tfrac{2}{3}\Pi(b^3 - R^3) = V^2 R^4 \rho\left\{\dfrac{1}{R} - \dfrac{1}{(R^3 + c^3)^{\frac{1}{3}}}\right\}.$

34. The determination of the motion of a liquid in a vessel of any given shape is one of great difficulty, and the solution has been effected in only a comparatively few number of cases. If, however, liquid is allowed to flow out of a vessel, the inclinations of whose sides to the vertical are small, an approximate solution may be obtained by neglecting the horizontal velocity of the liquid. This method of dealing with the problem is called the hypothesis of parallel sections.

TORRICELLI'S THEOREM.

Let us suppose that the vessel is kept full, and the liquid is allowed to escape by a small orifice at P. Let h be the distance of P below the free surface, and z that of any element of liquid. Since the motion is steady, the equation for the pressure will be

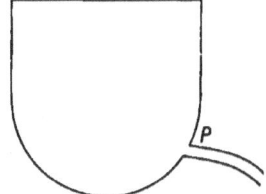

$$\frac{p}{\rho} - gz + \tfrac{1}{2}v^2 = C.$$

Now if the orifice be small in comparison with the area of the top of the vessel, the velocity at the free surface will be so small that it may be neglected; hence if Π be the atmospheric pressure, when $z = 0$, $p = \Pi$, $v = 0$ and therefore $C = \Pi/\rho$. At the orifice $p = \Pi$, $z = h$, whence the velocity of efflux is

$$v = \sqrt{2gh},$$

and is therefore the same as that acquired by a body falling from rest, through a height equal to the depth of the orifice below the upper surface of the liquid. This result is called *Torricelli's Theorem*.

35. Let us in the next place suppose that the vessel is a surface of revolution, which has a finite horizontal aperture, and which is kept full [1].

Let A be the area of the top AB of the vessel, U the velocity of the liquid there; let K, u; Z, v be similar quantities for the aperture CD, and a section ab whose depth below AB is z: also let h be the depth of CD below AB.

The conditions of continuity require that

$$AU = Ku = Zv,$$

and since the horizontal motion is neglected, the equation for the pressure is

$$\frac{1}{\rho}\frac{dp}{dz} = g - \frac{dv}{dt} - v\frac{dv}{dz}.$$

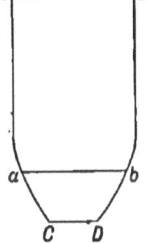

Now U and u are functions of t alone, whilst Z is a function of z only, whence

$$\frac{dv}{dt} = \frac{A}{Z}\frac{dU}{dt} = \frac{K}{Z}\frac{du}{dt},$$

whence
$$\frac{p}{\rho} = F(t) + gz - A\frac{dU}{dt}\int_0^z \frac{dz}{Z} - \tfrac{1}{2}v^2,$$

[1] Besant's *Hydromechanics*.

when $z = 0, p = \Pi, v = U$, therefore
$$\frac{\Pi}{\rho} = F(t) - \tfrac{1}{2}U^2,$$
when $z = h, p = \Pi, v = u$, whence if $a = \int_0^h Z^{-1}dz$,
$$\frac{\Pi}{\rho} = F(t) + gh - Aa\dot{U} - \tfrac{1}{2}u^2,$$
whence
$$Aa\dot{U} = gh + \tfrac{1}{2}(U^2 - u^2)$$
$$= gh + \tfrac{1}{2}U^2\left(1 - \frac{A^2}{K^2}\right).$$

Putting $(A/K)^2 - 1 = 2m$, $2\sqrt{ghm} = a\alpha$, and integrating, we obtain
$$U = \sqrt{\frac{gh}{m}} \frac{C - \epsilon^{-\alpha t}}{C + \epsilon^{-\alpha t}},$$
where C is the constant of integration. Now initially $U = 0$ since the motion is supposed to begin from rest, therefore $C = 1$, and we obtain
$$U = \sqrt{\frac{gh}{m}} \tanh \tfrac{1}{2}\alpha t$$
$$= \sqrt{\frac{gh}{m}} \tanh t \sqrt{ghm}/a.$$

The velocity of efflux is
$$u = \sqrt{(1 + 2m)\frac{gh}{m}} \tanh t \sqrt{ghm}/a.$$

After a very long time has elapsed $\tanh t\sqrt{ghm}/a$ becomes equal to unity, and if K be very small compared with A, $m = \infty$, and we obtain Torricelli's Theorem
$$u = \sqrt{2gh}.$$

The Vena Contracta.

36. When a jet of fluid escapes from a small hole in the bottom of a cistern, it is found that the area of the jet is less than the area of the hole; so that if σ be the area of the hole and σ' that of the jet, the ratio σ'/σ, which is called the *coefficient of contraction* of the jet, is always less than unity. We shall now show that this ratio must always be greater than $\tfrac{1}{2}$.

THE VENA CONTRACTA.

We shall suppose for simplicity, that no forces are in action, and that the jet escapes in vacuo; we shall also suppose that the upper surface of the liquid is subjected to a pressure p.

If the hole were absent, the pressure would be equal to p throughout the vessel, and therefore since the hole is small, the pressure may be taken to be sensibly equal to p except just in the neighbourhood of the hole, where it is zero.

If σ'' be the area of the cistern, v'' the velocity of the liquid across any section which is at some distance from the hole, the momentum which flows in across this section per unit of time is $\rho\sigma''v''^2$ and the momentum which flows out of the hole is $\rho\sigma'v'^2$; whence by the principle stated at the end of § 15

$$\rho\sigma''v''^2 - \rho\sigma'v'^2 = p\sigma'' - p(\sigma'' - \sigma) = p\sigma.$$

But since the pressure is zero at the hole

$$p/\rho - \tfrac{1}{2}v''^2 = -\tfrac{1}{2}v'^2.$$

Also the equation of continuity is

$$\sigma''v'' = \sigma'v',$$

whence eliminating p, v', v'' we obtain

$$\frac{2}{\sigma} = \frac{1}{\sigma'} + \frac{1}{\sigma''},$$

which shows that the coefficient of contraction is greater than $\tfrac{1}{2}$.

The quantity of liquid which flows out of the vessel per unit of time is therefore $\rho\sigma'v'$. Now if σ is small compared with σ'', we may neglect σ''^{-1}, and therefore $\sigma' = \tfrac{1}{2}\sigma$; hence the discharge is equal to

$$\tfrac{1}{2}\rho\sigma v',$$

where v' is the velocity of efflux.

Giffard's Injector[1].

37. If we suppose fluid of density ρ to escape through a small hole, from a large closed vessel in which the pressure is p at points where the motion is insensible, into an open space in which the pressure is Π, then if q be the velocity of efflux,

$$\Pi + \tfrac{1}{2}\rho q^2 = C, \quad p = C;$$

whence
$$q = \sqrt{\{2(p - \Pi)/\rho\}}.$$

---

[1] Greenhill, Art. Hydromechanics. *Encyc. Brit.*

If A be the sectional area of the jet at the vena contracta, the quantity of fluid which escapes per unit of time, is
$$\tfrac{1}{2} A \rho q = A \{2\rho (p - \Pi)\}.$$
The momentum per unit of time, is
$$A \rho q^2 = 2A (p - \Pi).$$
The energy per unit of time, is
$$\tfrac{1}{2} A \rho q^3 = A (p - \Pi)^{\frac{3}{2}} \sqrt{(2/\rho)}.$$

In Giffard's Injector, a jet of steam issuing by a pipe from the upper part of the boiler, is directed at an equal pipe leading back into the lower part of the boiler, the jet being kept constantly just surrounded with water. Now if we assume that the velocity of the steam jet, is equal to the velocity at which the water flows into the pipe leading to the lower part of the boiler, which must be very nearly true; it follows from the preceding equations that

$$\frac{\text{velocity of steam jet}}{\text{velocity of water jet}} = \sqrt{\frac{\rho}{\sigma}},$$

$$\frac{\text{quantity of steam jet}}{\text{quantity of water jet}} = \sqrt{\frac{\sigma}{\rho}},$$

$$\frac{\text{momentum of steam jet}}{\text{momentum of water jet}} = 1,$$

$$\frac{\text{energy of steam jet}}{\text{energy of water jet}} = \sqrt{\frac{\rho}{\sigma}},$$

where σ is the density of the steam jet.

If the steam and water jets were directed at each other with a small interval between them, the superior energy and equal momentum of the steam jet would overcome the water jet, and steam would be driven back into the boiler. But the steam jet without losing its momentum, is capable of being mixed with water to such an extent, as to become a condensed water jet moving with the velocity of the water jet, and still entering the boiler, a valve preventing the reversal of the motion. Consequently the amount of water carried into the boiler per unit of time, will theoretically at most be the difference between the quantities which would escape by the water and steam jets, and therefore
$$= A (p - \Pi)^{\frac{1}{2}} (\sqrt{2\rho} - \sqrt{2\sigma});$$
and therefore the efficiency of the jet, i.e. the ratio of the quantity of water pumped in, to the quantity of steam used, will be
$$\sqrt{\frac{\rho}{\sigma}} - 1.$$

EXAMPLES.

1. Find the equation of continuity in a form suitable for air in a tube, and prove that if the density be $f(at-x)$ where t is the time and x the distance from one end of a uniform tube, the velocity is
$$\frac{af(at-x)+(V-a)f(at)}{f(at-x)},$$
where V is the velocity at that end of the tube.

2. If the motion of a liquid be in two dimensions, prove that if at any instant the velocity be everywhere the same in magnitude, it is so in direction.

3. If every particle of a fluid move in the surface of a sphere, prove that the equation of continuity is
$$\frac{d\rho}{dt}\cos\theta + \frac{d}{d\theta}(\rho\omega\cos\theta) + \frac{d}{d\phi}(\rho\omega'\cos\theta) = 0,$$
where ρ is the density, θ and ϕ the latitude and longitude of any element, and ω, ω' the angular velocities of the element in latitude and longitude respectively.

4. Fluid is moving in a fine tube of variable section κ, prove that the equation of continuity is
$$\frac{d}{dt}(\kappa\rho) + \frac{d}{ds}(\kappa\rho v) = 0,$$
where v is the velocity at the point s.

5. If $F(x, y, z, t)$ is the equation of a moving surface, the velocity of the surface normal to itself is
$$-\frac{1}{R}\frac{dF}{dt} \text{ where } R^2 = (dF/dx)^2 + (dF/dy)^2 + (dF/dz)^2.$$

6. If x, y and z are given functions of a, b, c and t, where a, b and c are constants for any particular element of fluid, and if u, v and w are the values of $\dot{x}$, $\dot{y}$, $\dot{z}$ when a, b, c are eliminated, prove analytically that
$$\frac{d^2x}{dt^2} = \frac{du}{dt} + u\frac{du}{dx} + v\frac{du}{dy} + w\frac{du}{dz}.$$

32   EQUATIONS OF MOTION OF A PERFECT FLUID.

7. If the lines of flow of a fluid lie on the surfaces of coaxial cones having the same vertex, prove that the equation of continuity is

$$r\frac{d\rho}{dt} + r\frac{d}{dr}(u\rho) + 2\rho u + \operatorname{cosec}\theta \frac{d}{d\phi}(\rho v) = 0.$$

8. Show that

$$x^2/(akt^2)^2 + kt^2\{(y/b)^2 + (z/c)^2\} = 1$$

is a possible form of the bounding surface at time t of a liquid.

9. A fine tube whose section k is a function of its length s, in the form of a closed plane curve of area A filled with ice, is moved in any manner. When the component angular velocity of the tube about a normal to its plane is Ω, the ice melts without change of volume. Prove that the velocity of the liquid relatively to the tube at a point where the section is K, at any subsequent time when ω is the angular velocity is

$$\frac{2Ac}{K}(\Omega - \omega),$$

where $1/c = \int k^{-1}ds$, the integral being taken once round the tube.

10. A centre of force attracting inversely as the square of the distance, is at the centre of a spherical cavity within an infinite mass of liquid, the pressure on which at an infinite distance is ϖ, and is such that the work done by this pressure on a unit of area through a unit of length, is one half the work done by the attractive force on a unit of volume of the liquid from infinity to the initial boundary of the cavity; prove that the time of filling up the cavity will be

$$\pi a \sqrt{\frac{\rho}{\varpi}} \left\{ 2 - \left(\frac{3}{2}\right)^{\frac{1}{2}} \right\},$$

a being the initial radius of the cavity, and ρ the density of the liquid.

11. A solid sphere of radius a is surrounded by a mass of liquid whose volume is $4\pi c^3/3$, and its centre is a centre of attractive force varying directly as the square of the distance. If the solid sphere be suddenly annihilated, show that the velocity of the inner surface when its radius is x, is given by

$$\dot{x}^2 x^3 \{(x^3 + c^3)^{\frac{1}{3}} - x\} = \left(\frac{2\Pi}{3\rho} + \frac{2}{9}\mu c^2\right)(a^3 - x^3)(c^3 + x^3)^{\frac{1}{3}},$$

where ρ is the density, Π the external pressure and μ the absolute force.

12. Prove that if ϖ be the impulsive pressure, ϕ, ϕ' the velocity potentials immediately before and after an impulse acts, V the potential of the impulses,

$$\varpi + \rho V + \rho(\phi' - \phi) = \text{const.}$$

13. The surface of a vessel consists of two equal right cones, height $2c$, with coincident bases; it is fixed with its axis vertical and filled with water to half way up the axis of the upper cone, the air above this level being initially at atmospheric pressure and the vessel closed. The water flows out of the vessel from a ring of apertures on the level of bisection of the axis of the lower cone. On the hypothesis of parallel sections, obtain a differential equation for the velocity of efflux, while the free surface is above the midway point, and show that one equation to find its maximum value in this stage is

$$u^2[1 - \{c/(2c-x)\}^4] - 2g(c+x) = 2\Pi[\{c/(2c-x)\}^3 - 1]\rho^{-1},$$

where x = height of surface above midway point.

14. If the motion of a homogeneous liquid be given by a single valued velocity potential, prove that the angular momentum of any spherical portion of the liquid about its centre is always zero.

15. Homogeneous liquid is moving so that

$$u = \gamma x + \alpha y, \quad v = \beta x - \gamma y, \quad w = 0,$$

and a long cylindrical portion whose section is small, and whose axis is parallel to the axis of z, is solidified and the rest of the liquid destroyed. Prove that the initial angular velocity of the cylinder is

$$\frac{B\beta - A\alpha - 2F\gamma}{A + B},$$

where A, B, F are the moments and products of inertia of the section of the cylinder about the axes.

16. Fluid is contained within a sphere of small radius; prove that the momentum of the mass in the direction of the axis of x is greater than it would be if the whole were moving with the velocity at the centre by

$$\frac{Ma^2}{5\rho}\left\{\rho_x u_x + \rho_y u_y + \rho_z u_z + \tfrac{1}{2}\rho \nabla^2 u\right\},$$

where $\rho_x = d\rho/dx$ &c.

B. H.

34 EQUATIONS OF MOTION OF A PERFECT FLUID.

17. The motion of a liquid is in two dimensions, and there is a constant source at one point A in the liquid and an equal sink at another point B; find the form of the stream lines, and prove that the velocity at a point P varies as $(AP.BP)^{-1}$, the plane of the motion being unlimited.

If the liquid is bounded by the planes $x = 0$, $x = a$, $y = 0$, $y = a$, and if the source is at the point $(0, a)$ and the sink at $(a, 0)$, find an expression for the velocity potential.

18. The boundary of a liquid consists of an infinite plane having a hemispherical boss, whose radius is a and centre O. A doublet of unit strength is situated at a point S, whose axis coincides with OS, where OS is perpendicular to the plane. P is any point on the plane, $OP = y$, $OS = f$. Prove that the velocity of the liquid at P is

$$6fy\left\{\frac{a^3}{(a^4+f^2y^2)^{\frac{3}{2}}} - \frac{1}{(f^2+y^2)^{\frac{3}{2}}}\right\}.$$

19. Prove that

$$\phi = f(t)\{(r^2+a^2-2az)^{-\frac{1}{2}} + (r^2+a^2+2az)^{-\frac{1}{2}} - r^{-1}\} + \psi(t)$$

is the velocity potential of a liquid, and interpret it. Find the surfaces of equal pressure if gravity in the negative direction of the axis of z be the only force acting.

20. Liquid enters a right circular cylindrical vessel by a supply pipe at the centre O, and escapes by a pipe at a point A in the circumference; show that the velocity at any point P is proportional to $PB/PA.PO$, where B is the other end of the diameter AO. The vessel is supposed so shallow that the motion is in two dimensions.

CHAPTER II.

MOTION OF CYLINDERS AND SPHERES IN AN INFINITE LIQUID.

38. THE present chapter will be devoted to the consideration of certain problems of two-dimensional motion, and we shall also discuss the motion of a sphere in an infinite liquid.

If a right circular cylinder is moving in a liquid, the pressure of the liquid at any point of the cylinder passes through its axis, and therefore the resultant pressure of the liquid on the cylinder reduces to a single force, which can be calculated as soon as the pressure has been determined. Now the pressure at any point of the liquid, is found by means of the equation

$$V + p/\rho + \dot{\phi} + \tfrac{1}{2}q^2 = C,$$

and therefore p can be determined as soon as the velocity potential is known. Hence the first step towards the solution of problems of this character, is to find the velocity potential.

If the cylinder is not circular, the resultant pressure of the liquid upon its surface will usually be reducible to a single force and a couple, and the problem becomes more complicated. The motion of cylinders which are not circular, can be most conveniently treated by means of the dynamical methods explained in the next chapter. In the present chapter, we shall show how to find the motion of an infinite liquid, in which cylinders of certain given forms are moving, and we shall also work out the solution of certain special problems relating to the motion of circular cylinders and spheres.

39. If the liquid be at rest, and a cylinder of any given form be set in motion in any manner, the subsequent motion of the liquid will be irrotational and acyclic, and is therefore completely determined by means of a velocity potential. It is however more convenient to employ Earnshaw's current function ψ. This function, when the motion is irrotational, satisfies the equation

$$\frac{d^2\psi}{dx^2} + \frac{d^2\psi}{dy^2} = 0 \ldots\ldots\ldots\ldots\ldots\ldots(1)$$

at all points of the liquid.

The integral[1] of this equation is

$$\psi = f(x + \iota y) + F(x - \iota y) \ldots\ldots\ldots\ldots(2),$$

also
$$u = \frac{d\psi}{dy}, \quad v = -\frac{d\psi}{dx} \ldots\ldots\ldots\ldots\ldots(3).$$

We must now consider the boundary conditions to be satisfied by ψ.

If the liquid is at rest at infinity (which will usually be the case), $d\psi/dx$ and $d\psi/dy$ must vanish at infinity. If any portions of the boundary consist of fixed surfaces, the normal component of the velocity must vanish at such fixed boundaries, and therefore the fixed boundaries must coincide with a stream line. This requires that $\psi = $ const. at all points of fixed boundaries.

When the cylindrical boundary is in motion, the component velocity of the liquid along the normal, must be equal to the component velocity of the cylinder in the same direction.

(i) Let the cylinder be moving with velocity U parallel to the axis of x, and let θ be the angle which the normal to the cylinder makes with this axis; then at the surface

$$u \cos\theta + v \sin\theta = U \cos\theta.$$

Now $\cos\theta = dy/ds$; $\sin\theta = -dx/ds$; therefore by (3)

$$\frac{d\psi}{ds} = U \frac{dy}{ds}.$$

Integrating along the boundary, we obtain

$$\psi = Uy + A \ldots\ldots\ldots\ldots\ldots\ldots(4),$$

where A is a constant.

---

[1] The easiest way of showing that (2) is a solution of (1), is to differentiate the right-hand side of (2) twice with respect to x, and twice with respect to y and add. Since the result is zero, this shows that (2) *satisfies* (1); also since (2) contains *two* arbitrary *functions*, it is the most general solution that can be obtained.

BOUNDARY CONDITIONS. 37

(ii) If the cylinder be moving with velocity V parallel to the axis of y, the surface condition in the same manner can be shown to be

$$\psi = - Vx + B \quad \ldots\ldots\ldots\ldots\ldots\ldots\ldots\ldots\ldots (5).$$

(iii) Let the cylinder be rotating with angular velocity ω; then at the surface

$$u \cos\theta + v \sin\theta = - \omega y \cos\theta + \omega x \sin\theta,$$

or
$$\frac{d\psi}{ds} = - \omega r \frac{dr}{ds}.$$

Therefore $\quad \psi = -\tfrac{1}{2}\omega r^2 + C \ldots\ldots\ldots\ldots\ldots\ldots\ldots (6),$

where $r = (x^2 + y^2)^{\frac{1}{2}}$.

When there are any number of moving cylinders in the liquid, conditions (4), (5) and (6) must be satisfied at the surfaces of each of the moving cylinders.

In addition to the surface conditions, ψ must satisfy the following conditions at every point of space occupied by the liquid; viz. ψ must be a function which is a solution of Laplace's equation (1), and which together with its first derivatives must be finite and continuous at every point of the liquid.

If we take any solution of (1), and substitute its value in (4), (5) or (6), we shall in many cases be able to determine the current function due to the motion of a cylinder, whose cross section is some curve, in one of the three prescribed manners.

In most of the applications which follow ψ will be of the form

$$\psi = f(x + \iota y) + f(x - \iota y) \ldots\ldots\ldots\ldots\ldots (7),$$

also by (3) $\quad \dfrac{d\phi}{dx} = \dfrac{d\psi}{dy}, \quad \dfrac{d\phi}{dy} = - \dfrac{d\psi}{dx} \ldots\ldots\ldots\ldots\ldots (8).$

From these equations we see that

$$\phi + \iota \psi = 2\iota f(x + \iota y) \ldots\ldots\ldots\ldots\ldots\ldots (9),$$

and therefore when ψ is known, ϕ can be found by equating the real and imaginary parts of (9).

Motion of a Circular Cylinder.

40. Let
$$\psi = -\tfrac{1}{2} V a^2 \left(\frac{1}{x+\iota y} + \frac{1}{x-\iota y} \right).$$

Transforming to polar coordinates, and using De Moivre's theorem, we obtain
$$\psi = - V a^2 x / r^2 \ldots\ldots\ldots(10).$$

When $r = a$, $\psi = - V x$; equation (10) consequently determines the current function, when a circular cylinder of radius a is moving parallel to the axis of y, in an infinite liquid with velocity V.

By (9) the velocity potential is
$$\phi = - V a^2 y / r^2 \ldots\ldots\ldots(11).$$

41. Let us now suppose that the cylinder is of finite length unity, and that the liquid is bounded by two vertical parallel planes, which are perpendicular to the axis of the cylinder.

In order to find the motion, when the cylinder is descending vertically under the action of gravity, let β be the distance of the axis of the cylinder at time t from some fixed point in its line of motion which we shall choose as the origin, and let (x, y) be the coordinates of any point of the liquid referred to the fixed origin, the axis of y being measured vertically downwards; also let (r, θ) be polar coordinates of the same point referred to the axis of the cylinder as origin. By (11)
$$\phi = - \frac{V a^2}{r} \sin \theta = - \frac{V a^2 (y - \beta)}{x^2 + (y - \beta)^2},$$

and therefore since $d\beta/dt = V$,
$$\dot\phi = - \frac{a^2 \dot V}{r} \sin \theta + \frac{a^2 V^2}{r^2} - \frac{2 a^2 V^2}{r^2} \sin^2 \theta,$$

and therefore at the surface, where $r = a$,
$$\dot\phi = - a \dot V \sin \theta + V^2 \cos 2\theta.$$

Also
$$q^2 = \left(\frac{d\phi}{dr}\right)^2 + \left(\frac{1}{r} \frac{d\phi}{d\theta}\right)^2;$$

therefore when $r = a$,
$$q^2 = V^2.$$

Whence
$$p/\rho = a\dot{V}\sin\theta - V^2\cos 2\theta - \tfrac{1}{2}V^2 + g(\beta + a\sin\theta) + C\ldots(12).$$

The horizontal resultant of the pressure is evidently zero; the vertical resultant is
$$Y = -a\int_0^{2\pi} p\sin\theta\,d\theta.$$

Substituting the value of p from (12) and integrating, we obtain
$$Y = -\pi\rho a^2(\dot{V} + g).$$

Hence if σ be the density of the cylinder, the equation of motion is
$$\pi\sigma a^2\dot{V} = Y + \pi\sigma g a^2,$$
or
$$(\sigma + \rho)\dot{V} = (\sigma - \rho)g\ldots\ldots\ldots\ldots\ldots(13).$$

Integrating this equation, we obtain
$$V = v + \frac{(\sigma - \rho)gt}{(\sigma + \rho)},$$
where v is the initial velocity measured vertically downwards.

We therefore see that the cylinder will move in a vertical straight line, with a constant acceleration which is equal to
$$g(\sigma - \rho)/(\sigma + \rho).$$

In order to pass to the case in which there are no forces in action, we must put $g = 0$, in which case V remains constant and equal to its initial value. It thus appears that the only effect of the liquid is, to produce an apparent increase in the inertia of the cylinder, which is equal to the mass of the liquid displaced.

By combining these two results, we see that if the cylinder be projected in any manner under the action of gravity, it will describe a parabola with vertical acceleration $g(\sigma - \rho)/(\sigma + \rho)$.

It is well known that if a solid body be projected in a liquid of unlimited extent, and no impressed forces are in action, it will not continue to move with constant velocity, but will gradually come to rest. One reason of this discrepancy is, that we have proceeded upon the supposition that the liquid is frictionless, whereas all liquids with which we are acquainted are more or less viscous, which produces a gradual conversion of kinetic energy into heat. We shall consider this question more fully when discussing the motion of a sphere.

42. The motion of a cylinder in a liquid, which is bounded by a fixed external cylinder, is a problem of considerable difficulty. If however the cylinders are initially concentric, the initial motion can easily be found; and this problem will afford an example of the use of the velocity potential.

The velocity potential, as we know, satisfies Laplace's equation, which when transformed[1] into polar coordinates becomes

$$\frac{d^2\phi}{dr^2} + \frac{1}{r}\frac{d\phi}{dr} + \frac{1}{r^2}\frac{d^2\phi}{d\theta^2} = 0 \quad\ldots\ldots\ldots\ldots\ldots\ldots (14).$$

Let us endeavour to satisfy this equation by assuming $\phi = F(r)\,\epsilon^{\iota n\theta}$; this will be possible, provided

$$\frac{d^2F}{dr^2} + \frac{1}{r}\frac{dF}{dr} - \frac{n^2 F}{r^2} = 0.$$

Assuming $F = r^m$, the equation reduces to $m^2 - n^2 = 0$, whence $m = \pm n$, and therefore the required solution is

$$\phi = (Ar^n + Br^{-n})\,\epsilon^{\iota n\theta}\ldots\ldots\ldots\ldots\ldots\ldots(15).$$

In this solution n may have any value whatever, and the real and imaginary parts of the above expression will be independent solutions of (14).

Let us now suppose that the radius of the outer cylinder, which is supposed to be fixed, is c; and let the inner one be started with velocity U.

Since the velocity of the liquid at the surface of the outer cylinder must be wholly tangential, the boundary condition is

$$\frac{d\phi}{dr} = 0, \quad\text{when}\quad r = c\ldots\ldots\ldots\ldots\ldots\ldots(16).$$

At the surface of the inner cylinder, which is moving with velocity U, the component velocities of the cylinder and liquid along the radius, must be equal; whence the boundary condition at the inner cylinder is

$$\frac{d\phi}{dr} = U\cos\theta, \quad\text{when}\quad r = a\ldots\ldots\ldots\ldots\ldots\ldots(17),$$

θ being measured from the direction of U.

If in (15) we put $n = 1$, the function

$$\phi = (Ar + B/r)\cos\theta\ldots\ldots\ldots\ldots\ldots\ldots(18)$$

---

[1] This transformation can be most easily effected, by forming the equation of continuity in polar coordinates.

LEMNISCATE OF BERNOULLI. 41

is a solution of Laplace's equation; if therefore we can determine A and B so as to satisfy (16) and (17), the problem will be solved.

Substituting from (18) in (16) we obtain
$$Ac^2 - B = 0.$$
Substituting in (17) we obtain
$$Aa^2 - B = Ua^2.$$
Solving and substituting in (18) we obtain
$$\phi = -\frac{Ua^2}{c^2 - a^2}\left(r + \frac{c^2}{r}\right)\cos\theta.$$

If we put $c = \infty$, we fall back on our previous result of a cylinder moving in an infinite liquid.

We can now determine the impulsive force which must be applied to the inner cylinder, in order to start it with velocity U.

By § 23, equation (34), it follows that if a liquid which is at rest be set in motion by means of an impulse, and ϕ be the velocity potential of the initial motion, the impulsive pressure at any point of the liquid is equal to $-\rho\phi$.

Hence if M be the mass of the cylinder, F the impulse, the equation of motion is
$$MU = F - \int_0^{2\pi} p\cos\theta\, d\theta$$
$$= F + \rho a \int_0^{2\pi} \phi \cos\theta\, d\theta$$
$$= F - \frac{U\pi\rho a^2 (c^2 + a^2)}{c^2 - a^2},$$
whence since $M = \pi\sigma a^2$
$$F = \left\{1 + \frac{\rho}{\sigma}\frac{(c^2 + a^2)}{(c^2 - a^2)}\right\} MU.$$

The Lemniscate of Bernoulli.

43. The lemniscate of Bernoulli is a quartic curve whose equation in Cartesian coordinates is $(x^2 + y^2)^2 = 2c^2(x^2 - y^2)$, or in polar coordinates $r^2 = 2c^2 \cos 2\theta$. In order to find the current function, when a cylinder, whose cross section is this curve, is moving parallel to x in an infinite liquid, let us put $u = x + \iota y$, $v = x - \iota y$, and assume
$$\psi_z = \tfrac{1}{2}Uc\iota\left\{\frac{u}{(u^2 - c^2)^{\frac{1}{2}}} - \frac{v}{(v^2 - c^2)^{\frac{1}{2}}}\right\}\quad\ldots\ldots\ldots(19).$$

Now $u^2 - c^2 = r^2(\cos 2\theta + \iota \sin 2\theta) - c^2$;

whence at the surface where $r^2 = 2c^2 \cos 2\theta$, the right-hand side becomes

$$2c^2 \cos^2 2\theta + \iota c^2 \sin 4\theta - c^2 = c^2 (\cos 2\theta + \iota \sin 2\theta)^2;$$

whence $\dfrac{u}{(u^2 - c^2)^{\frac{1}{2}}} = \dfrac{r(\cos\theta + \iota \sin\theta)}{c(\cos 2\theta + \iota \sin 2\theta)} = r(\cos\theta - \iota \sin\theta)/c.$

Therefore at the surface

$$\psi_x = Ur \sin\theta = Uy.$$

The value of ψ_x given by (19), is therefore the current function due to the motion parallel to x with velocity U, of a cylinder whose cross section is a lemniscate of Bernoulli.

If we put $\psi_y = -\tfrac{1}{2} Vc \left\{ \dfrac{u}{(u^2 - c^2)^{\frac{1}{2}}} + \dfrac{v}{(v^2 - c^2)^{\frac{1}{2}}} \right\},$

$$\psi_3 = -\tfrac{1}{2} \omega c^3 \left\{ \dfrac{1}{(u^2 - c^2)^{\frac{1}{2}}} + \dfrac{1}{(v^2 - c^2)^{\frac{1}{2}}} \right\},$$

it can be shown in a similar manner, that ψ_y is the current function, when a cylinder of this form is moving parallel to y with velocity V; and that ψ_3 is the current function, when the cylinder is rotating with angular velocity ω about its axis.

If the cross section be the cardioid $r = 2c(1 + \cos\theta)$, the values of ψ_x and ψ_y can be obtained by writing $(u^{\frac{1}{2}} - c^{\frac{1}{2}})^2$, $(v^{\frac{1}{2}} - c^{\frac{1}{2}})^2$ for $(u^2 - c^2)^{\frac{1}{2}}$, $(v^2 - c^2)^{\frac{1}{2}}$ in the preceding formulae; but the value of ψ_3 cannot be so simply obtained. See *Quart. Jour.* vol. XX. p. 246.

An Equilateral Triangle.

44. The preceding methods may also be employed, to find the motion of a liquid, which is contained within certain cylindrical cavities, which are rotating about an axis.

Let $\psi = \tfrac{1}{2} A \{(x + \iota y)^3 + (x - \iota y)^3\}$
$= A(x^3 - 3xy^2) = Ar^3 \cos 3\theta.$

Substituting in (6), the boundary condition becomes

$$A(x^3 - 3xy^2) + \tfrac{1}{2}\omega(x^2 + y^2) = C \dots\dots\dots\dots(20).$$

If we choose the constants so that the straight line $x = a$, may form part of the boundary, we find

$$A = \dfrac{\omega}{6a}; \quad C = \dfrac{2\omega a^2}{3}.$$

Hence (20) splits up into the factors

$$(x-a);\ x+y\sqrt{3}+2a;\ x-y\sqrt{3}+2a.$$

The boundary therefore consists of three straight lines forming an equilateral triangle, whose centre of inertia is the origin.

Hence ψ is the current function due to liquid contained in an equilateral prism, which is rotating with angular velocity ω about an axis through the centre of inertia of its cross section. The values of ψ and ϕ, when cleared of imaginaries, are

$$\psi = \frac{\omega}{6a} r^3 \cos 3\theta, \quad \phi = \frac{\omega}{6a} r^3 \sin 3\theta.$$

An Elliptic Cylindrical Cavity.

45. Let $\quad \psi = \tfrac{1}{2}A\{(x+\iota y)^2 + (x-\iota y)^2\}$
$= A(x^2 - y^2).$

Substituting in (6) we find

$$A(x^2 - y^2) + \tfrac{1}{2}\omega(x^2 + y^2) = C.$$

Putting $\quad \dfrac{\omega + 2A}{2C} = \dfrac{1}{a^2};\ \dfrac{\omega - 2A}{2C} = \dfrac{1}{b^2},$

the equation of the boundary becomes

$$\frac{x^2}{a^2} + \frac{y^2}{b^2} = 1,$$

and $\quad \psi = -\dfrac{\omega(a^2 - b^2)}{2(a^2 + b^2)}(x^2 - y^2),$

ψ is therefore the current function due to the motion of liquid contained in an elliptic cylinder, which is rotating about its axis.

Elliptic Cylinder.

46. The problem of finding the motion of an elliptic cylinder in an infinite liquid, cannot be solved by such simple methods as the foregoing; in order to effect the solution we require to employ the method of Conjugate Functions.

Def. *If ξ and η are functions of x and y, such that*

$$\xi + \iota\eta = f(x + \iota y) \ \dotfill (21),$$

then ξ and η are called conjugate functions of x and y.

If we differentiate (21) first with respect to x, and afterwards

with respect to y, eliminate the arbitrary function, and then equate the real and imaginary parts, we shall obtain the equations
$$\frac{d\xi}{dx} = \frac{d\eta}{dy}, \quad \frac{d\xi}{dy} = -\frac{d\eta}{dx} \dots \dots \dots \dots \dots (22).$$

Comparing these equations with (8), we see that ϕ and ψ are conjugate functions of x and y.

From equations (22) we also see that
$$\frac{d\xi}{dx}\frac{d\eta}{dx} + \frac{d\xi}{dy}\frac{d\eta}{dy} = 0 \dots \dots \dots \dots \dots (23),$$
$$\nabla^2\xi = 0, \quad \nabla^2\eta = 0 \dots \dots \dots \dots \dots (24).$$

Equation (23) shows that the curves $\xi = $ const., $\eta = $ const., form an orthogonal system; and equations (24) show that ξ and η each satisfy Laplace's equation.

If ϕ and ψ are conjugate functions of x and y, and ξ and η are also conjugate functions of x and y, then ϕ and ψ are conjugate functions of ξ and η.

For $\quad\quad\quad \phi + \iota\psi = F(x + \iota y),$
and $\quad\quad\quad \xi + \iota\eta = f(x + \iota y),$
whence eliminating $x + \iota y$, we have
$$\phi + \iota\psi = \chi(\xi + \iota\eta).$$

From this proposition combined with (24), it follows that if the equation $\nabla^2\psi = 0$ be transformed by taking ξ and η as independent variables
$$\frac{d^2\psi}{d\xi^2} + \frac{d^2\psi}{d\eta^2} = 0 \dots \dots \dots \dots \dots (25).$$

47. We can now find the current function due to the motion of an elliptic cylinder.

Let $\quad x + \iota y = c \cos(\xi - \iota\eta)$
$\quad\quad\quad\quad = c \cos\xi \cosh\eta + \iota c \sin\xi \sinh\eta,$
then $\quad\quad\quad x = c \cos\xi \cosh\eta$
$\quad\quad\quad\quad y = c \sin\xi \sinh\eta \Big\} \dots \dots \dots \dots \dots (26),$

whence the curves $\eta = $ const., $\xi = $ const., represent a family of confocal ellipses and hyperbolas, the distance between the foci being $2c$.

If a and b be the semi-axes of the cross section of the elliptic cylinder $\eta = \beta$, then,
$$a = c \cosh\beta, \quad b = c \sinh\beta.$$

ELLIPTIC CYLINDER.   45

If β is exceedingly large, $\sinh\beta$ and $\cosh\beta$ both approximate to the value $\frac{1}{2}ce^\beta$; and therefore as the ellipse increases in size, it approximates to a circle whose radius is $\frac{1}{2}ce^\beta$.

It can be verified by trial, that (25) can be satisfied by a series of terms of the form $\epsilon^{-n\eta}(A_n \cos n\xi + B_n \sin n\xi)$; and if n be a positive quantity not less than unity, this is the proper form of ψ outside an elliptic cylinder, since it continually diminishes as η increases.

When the cylinder is moving parallel to its major axis with velocity U, let us assume
$$\psi_x = A\epsilon^{-\eta}\sin\xi.$$
Substituting in (4) we obtain
$$A\epsilon^{-\beta}\sin\xi = Uc\sinh\beta\sin\xi + C,$$
where $\eta = \beta$ is the equation of the cross section of the cylinder.

Since this equation is to be satisfied at every point of the boundary, we must have $C = 0$, $A = Uce^\beta \sinh\beta$; whence
$$\psi_x = Uc\epsilon^{-\eta+\beta}\sinh\beta\sin\xi\dots\dots\dots\dots(27).$$

When the cylinder is moving parallel to its minor axis with velocity V, it may be shown in the same manner that
$$\psi_y = -Vc\epsilon^{-\eta+\beta}\cosh\beta\cos\xi\dots\dots\dots\dots(28).$$

Lastly let us suppose that the cylinder is rotating with angular velocity ω about its axis. Then
$$x^2 + y^2 = c^2(\cos^2\xi\cosh^2\eta + \sin^2\xi\sinh^2\eta)$$
$$= \tfrac{1}{2}c^2(\cosh 2\eta + \cos 2\xi).$$

Let us therefore assume
$$\psi_3 = B\epsilon^{-2\eta}\cos 2\xi$$
Substituting in (6) we obtain
$$B\epsilon^{-2\beta}\cos 2\xi + \tfrac{1}{4}\omega c^2(\cosh 2\beta + \cos 2\xi) = C,$$
whence $B = -\tfrac{1}{4}\omega c^2 \epsilon^{2\beta}, \quad C = \tfrac{1}{4}\omega c^2 \cosh 2\beta,$
and therefore $\psi_3 = -\tfrac{1}{4}\omega c^2 \epsilon^{-2(\eta-\beta)}\cos 2\xi \dots\dots\dots\dots(29).$

48. If we suppose that $\beta = 0$, the ellipse degenerates into a straight line joining the foci, and (28) becomes
$$\psi_y = -Vc\epsilon^{-\eta}\cos\xi\dots\dots\dots\dots\dots(30).$$

It might therefore be supposed that (30) gives the value of the current function, due to a lamina of breadth $2c$, which moves with velocity V, perpendicularly to itself. This however is not the case, inasmuch as the velocity at the edges of the lamina becomes infinite, and therefore the solution fails. To prove this, we have

$$\frac{d\psi}{d\eta} = \frac{d\psi}{dx}\frac{dx}{d\eta} + \frac{d\psi}{dy}\frac{dy}{d\eta}$$

$$= c \sinh \eta \cos \xi \, \frac{d\psi}{dx} + c \cosh \eta \sin \xi \, \frac{d\psi}{dy},$$

and $$\frac{d\psi}{d\xi} = - c \cosh \eta \sin \xi \, \frac{d\psi}{dx} + c \sinh \eta \cos \xi \, \frac{d\psi}{dy},$$

whence squaring and adding, we obtain

$$c^2(\sinh^2 \eta \cos^2 \xi + \cosh^2 \eta \sin^2 \xi) q^2 = \left(\frac{d\psi}{d\eta}\right)^2 + \left(\frac{d\psi}{d\xi}\right)^2 = V^2 c^2 \epsilon^{-2\eta} \dots (31).$$

The coordinates of an edge are $x = \pm c$, $y = 0$; and therefore in the neighbourhood of an edge η and ξ are very small quantities; and therefore by (31) the velocity in the neighbourhood of an edge is

$$q = \frac{V}{(\eta^2 + \xi^2)^{\frac{1}{2}}},$$

which becomes infinite at the edge itself, where η and ξ are zero. It therefore follows that the pressure in the neighbourhood of an edge is negative, which is physically impossible.

Since the pressure is positive at a sufficient distance from the edge, there will be a surface of zero pressure dividing the regions of positive and negative pressures; and it might be thought that the interpretation of the formulae would be, that a hollow space exists in the liquid surrounding the edges, which is bounded by a surface of zero pressure. But the condition that a free surface should be a surface of zero (or constant) pressure, although a necessary one, is not sufficient; it is further necessary, that such a surface should be a surface of no flux, which satisfies the kinematical condition of a bounding surface § 12, equation (17); and it will be found on investigating the question, that no surface exists, which is a surface of zero (or constant) pressure, and at the same time satisfies the conditions of a bounding surface. The solution altogether fails in the case of a lamina.

When the velocity of the solid is constant and equal to V, the easiest way of dealing with a problem of this character, is to reverse the motion by supposing the solid at rest, and the liquid flowing past it, the velocity at infinity being equal to $-V$. The correct solution in the case of a lamina has been given by Kirchhoff[1], and he has shown that behind the lamina there is a region of *dead water*, i.e. water at rest, which is separated from the remainder of the liquid by two surfaces of discontinuity, which commence at the two edges of the lamina, and proceed to infinity in the direction in which the stream is flowing. Since the liquid on one side of this surface of discontinuity is at rest, its pressure is constant; and therefore since the motion is steady, the pressure, and therefore the velocity of the moving liquid, must be constant at every point of the surface of discontinuity. It may be added that a surface of discontinuity, is an imaginary surface described in the liquid, such that the tangential component of the velocity suddenly changes as we pass from one side of the surface to the other.

Motion of a Sphere.

49. The determination of the velocity potential, when a solid body of any given shape is moving in an infinite liquid, is one of great difficulty, and the only problem of the kind which has been completely worked out, is that of an ellipsoid, which of course includes a sphere as a particular case.

We shall however find it simpler in the case of a sphere, to solve the problem directly, which we shall proceed to do.

Let the sphere be moving along a straight line with velocity V, and let (r, θ, ω) be polar coordinates referred to the centre of the sphere as origin, and to the direction of motion as initial line. The conditions of symmetry show that ϕ must be a function of (r, θ) and not of ω, hence by § 7, equation (11), the equation of continuity is

$$\frac{d^2\phi}{dr^2} + \frac{2}{r}\frac{d\phi}{dr} + \frac{1}{r^2}\frac{d^2\phi}{d\theta^2} + \frac{1}{r^2}\cot\theta\,\frac{d\phi}{d\theta} = 0 \ldots\ldots\ldots (32).$$

The boundary condition, which expresses the fact, that the normal component of the velocity of the liquid at the surface of the sphere, is equal to the normal component of the velocity of the sphere itself, is

$$\frac{d\phi}{dr} = V\cos\theta \ldots\ldots\ldots\ldots\ldots\ldots (33).$$

Equation (33) suggests that ϕ must be of the form $F(r)\cos\theta$; we shall therefore try whether we can determine F so as to satisfy

[1] See also, Michell, On the theory of free stream lines, *Proc. Roy. Soc.* vol. XLVII. p. 129.

(32). Substituting this value of ϕ, we find that (32) will be satisfied, provided

$$\frac{d^2F}{dr^2} + \frac{2}{r}\frac{dF}{dr} - \frac{2F}{r^2} = 0 \quad\ldots\ldots\ldots\ldots\ldots\ldots (34).$$

To solve (34), assume $F = r^m$; whence on substitution we obtain

$$(m-1)(m+2) = 0;$$

which requires that $m = 1$ or -2.

A particular solution of (32) is therefore

$$\phi = (Ar + Br^{-2})\cos\theta.$$

Since the liquid is supposed to be at rest at infinity, $d\phi/dr = 0$ when $r = \infty$, and therefore $A = 0$. To find B, substitute in (33) and put $r = a$, and we find

$$A = -\tfrac{1}{2}Va^3,$$

whence $$\phi = -\frac{Va^3\cos\theta}{2r^2} \quad\ldots\ldots\ldots\ldots\ldots\ldots\ldots (35).$$

This is the expression for the velocity potential due to the motion of a sphere in an infinite liquid.

In order to determine the motion, when a sphere is descending vertically under the action of gravity, let γ be the distance of its centre at time t from some fixed point in its line of motion, which we shall choose as the origin; let the axis of z be measured vertically downwards and let x, y, z be the coordinates of any point of the liquid referred to the *fixed* origin.

By (35) $$\phi = -\frac{Va^3(z-\gamma)}{2\{x^2+y^2+(z-\gamma)^2\}^{\frac{3}{2}}},$$

and therefore since $\dot\gamma = V$,

$$\frac{d\phi}{dt} = -\frac{\dot{V}a^3\cos\theta}{2r^2} + \frac{V^2 a^3}{2r^3} - \frac{3V^2 a^3 \cos^2\theta}{2r^3};$$

and therefore at the surface where $r = a$,

$$\frac{d\phi}{dt} = -\tfrac{1}{2}\dot{V}a\cos\theta - \tfrac{1}{2}V^2(3\cos^2\theta - 1).$$

also $$q^2 = \left(\frac{d\phi}{dr}\right)^2 + \left(\frac{1}{r}\frac{d\phi}{d\theta}\right)^2$$

$$= V^2(\cos^2\theta + \tfrac{1}{4}\sin^2\theta),$$

and therefore

$$p/\rho = C + g(\gamma + a\cos\theta) + \tfrac{1}{2}\dot{V}a\cos\theta + \tfrac{1}{8}V^2(9\cos^2\theta - 5)\ldots(36).$$

If Z be the force due to the liquid pressure, which opposes the motion,

$$Z = \iint p \cos\theta \, dS$$

$$= 2\pi a^2 \int_0^\pi p \cos\theta \sin\theta \, d\theta$$

$$= \tfrac{4}{3}\pi\rho a^3 (\tfrac{1}{2}\dot{V} + g) \quad\ldots\ldots\ldots\ldots\ldots\ldots (37)$$

by (36). If therefore σ be the density of the sphere, the equation of motion is

$$\tfrac{4}{3}\pi\sigma a^3 \dot{V} = -\tfrac{4}{3}\pi\rho a^3 (\tfrac{1}{2}\dot{V} + g) + \tfrac{4}{3}\pi a^3 \sigma g$$

or $\qquad\qquad (\sigma + \tfrac{1}{2}\rho)\dot{V} = (\sigma - \rho)g \ldots\ldots\ldots\ldots\ldots\ldots(38).$

Hence the sphere descends with vertical acceleration

$$g(\sigma - \rho)/(\sigma + \tfrac{1}{2}\rho).$$

In order to pass to the case in which the sphere is projected with a given velocity and no forces are in action, we must put $g = 0$, and we see that $V = $ const. $=$ its initial value; hence the sphere continues to move with its velocity of projection, and the effect of the liquid is to produce an apparent increase in the inertia of the sphere, which is equal to half the mass of the liquid displaced.

It also follows that if the sphere be projected in any manner under the action of gravity, it will describe a parabola with vertical acceleration $g(\sigma - \rho)/(\sigma + \tfrac{1}{2}\rho)$.

50. Let us now suppose that the sphere is moving with constant velocity V under the action of no forces. The equation determining the pressure is

$$\frac{p}{\rho} = C + V\frac{d\phi}{dz} - \tfrac{1}{2}q^2.$$

Since $d\phi/dz$ and q vanish at infinity, it follows that $C = \Pi/\rho$, where Π is the pressure at infinity, whence

$$\frac{p}{\rho} = \frac{\Pi}{\rho} + V\frac{d\phi}{dz} - \tfrac{1}{2}q^2;$$

and therefore at the surface,

$$\frac{p}{\rho} = \frac{\Pi}{\rho} + \tfrac{1}{8}V^2(9\cos^2\theta - 5).$$

The right-hand side of this equation will be a minimum when $\theta = \tfrac{1}{2}\pi$, in which case it becomes $\Pi/\rho - \tfrac{5}{8}V^2$. Hence if

$$\Pi < \tfrac{5}{8}V^2\rho$$

50   MOTION OF CYLINDERS AND SPHERES.

the pressure will become negative within a certain region in the neighbourhood of the equator, and the solution fails. If Π be given, it is by no means clear what happens when V exceeds the critical value $(8\Pi/5\rho)^{\frac{1}{2}}$, and the problem is one which awaits solution.

51. In discussing the motion of a cylinder, we also found that if the solid were projected in a liquid and no forces were in action, the solid would continue to move in a straight line with its original velocity of projection; and we called attention to the fact that this result was contrary to experience; and that one reason of this discrepancy between theory and observation arose from the fact that all liquids are more or less viscous, the result of which is that the kinetic energy is gradually converted into heat. The motion of viscous fluids is beyond the scope of an elementary work such as the present, but a few remarks on the subject will not be out of place.

Let us suppose that fluid is moving in strata parallel to the plane xy, with a variable velocity U, which is parallel to the axis of x. Let U be the velocity of the stratum AB, $U + \delta U$ of the stratum CD, and let δz be the distance between AB and CD.

If the fluid were frictionless, the action between the fluid on either side of the plane AB, would be a hydrostatic pressure p, whose direction is perpendicular to this plane, and consequently no tangential action or shearing stress could exist; if however the fluid is viscous, the action between the fluid on either side of the plane AB, usually consists of an oblique pressure (or tension), and may therefore be resolved into a normal component *perpendicular* to the plane, and a tangential component *in* the plane.

The usual theory of viscosity supposes, that if F be the tangential stress on AB per unit of area, $F + \delta F$ the corresponding stress on CD, then the latter stress is proportional to the relative velocity of the two strata divided by the distance between the strata, so that

$$F + \delta F \propto \frac{U + \delta U - U}{\delta z},$$

whence proceeding to the limit

$$F \propto \frac{dU}{dz}.$$

We may therefore put $\quad F = \mu \dfrac{dU}{dz}$ ....(39), where μ is a constant. The constant μ is called the *coefficient of viscosity*; it is a numerical quantity whose value is different for different fluids, and also depends upon the temperature.

The coefficient of viscosity is a quantity which corresponds to the *rigidity* in the Theory of Elasticity. If a shearing stress F be applied parallel to the axis of x, and in a plane parallel to the plane xy, to an elastic solid, it is known from the Theory of Elasticity, that

$$F = n d\alpha/dz,$$

where α is the displacement parallel to x. Whence the ratio of the shearing stress F, to the shearing strain $d\alpha/dz$ produced by it, is equal to a constant n, which is called the *rigidity*. Now in the hydrodynamical theory of viscous fluids, dU/dz is equal to the rate at which shearing strain is produced by the shearing stress F; hence (39) asserts that the ratio of the shearing stress to the rate at which shearing strain is produced, is equal to a constant μ, which is called the coefficient of viscosity.

If the shearing stress F is applied in the plane $z = c$, and if $U = uz/c$, (39) becomes

$$F = \mu u/c \quad\quad\quad\quad\quad (40),$$

where u is the velocity of the fluid in the plane $z = c$. Hence if $u = 1$, and $c = 1$, then $F = \mu$. We may therefore define the coefficient of viscosity as follows[1].

The coefficient of viscosity is equal to the tangential force per unit of area, on either of two parallel planes at the unit of distance apart, one of which is fixed, whilst the other moves with the unit of velocity, the space between being filled with the viscous fluid.

Equation (40) shows that the dimensions of μ are $[ML^{-1}T^{-1}]$.

If we put $\nu = \mu/\rho$, where ρ is the density, the quantity ν is called the *kinematic coefficient of viscosity*. The dimensions of ν are $[L^2 T^{-1}]$.

The equations of motion of a viscous fluid are known, and the motion of a sphere which is descending under the action of gravity in a slightly viscous *liquid*, such as water, has been worked out by myself; and I have shown that if the sphere be initially projected downwards with velocity V, its velocity at any subsequent time will be approximately given by the equation

[1] Maxwell's *Heat*, p. 298, fourth edition.

$$v = f(1 - e^{-\lambda t})/\lambda + Ve^{-\lambda t}$$

where
$$f = \frac{(\sigma - \rho)g}{\sigma + \tfrac{1}{2}\rho}, \qquad \lambda = \frac{9\mu}{a^2(2\sigma + \rho)}.$$

If no forces are in action, the velocity at time t is

$$v = Ve^{-\lambda t},$$

which shows that the velocity diminishes with the time.

If the liquid were frictionless, μ would be zero, and we should fall back on our previous results.

52. The resistance which a ship experiences in moving through water, is principally due to the following three causes, (i) viscosity, (ii) skin friction, (iii) wave resistance.

The first has been already considered, and since the coefficient of viscosity of water is a small quantity, viz. ·014 dynes per square centimetre in C. G. S. units, the temperature being $24\cdot 5°$ C.; it would appear that the effect of viscosity is small.

The second cause is due to the friction between the sides of the vessel and the water in contact with it, and in the opinion of Mr Froude[1], is the principal cause of the resistance experienced by ships.

The third cause is due to the fact, that when a ship is in motion on a river or in the open sea, waves are continually being generated, which require an expenditure of energy for their production, and this is necessarily supplied by the mechanical power, which is employed to propel the ship. If therefore the ship were set in motion in a frictionless liquid and left to itself, the initial energy would gradually be dissipated in forming waves, and would be carried away by them, so that this cause alone would ultimately bring the ship to rest.

53. Having made this digression upon viscosity and resistance, we must return to the subject of this chapter.

Let us suppose that a spherical pendulum, which is surrounded with liquid, is performing small oscillations in a vertical plane.

Let $l - a$ be the length of the pendulum rod, whose mass we shall suppose small enough to be neglected, and let M' be the mass of the liquid displaced.

[1] On Stream lines in relation to the resistance of ships. *Nature*, Vol. III.

MOTION IN A SPHERICAL ENVELOP. 53

If V be the horizontal velocity of the pendulum, it follows from (37), that the horizontal and vertical forces due to the pressure of the liquid are
$$X = \tfrac{1}{2} M'\dot{V}, \quad Y = M'g.$$
Now $V = a\dot{\theta}$, whence taking moments about the point of suspension, we obtain
$$M(\tfrac{2}{5}a^2 + l^2)\ddot{\theta} = -Mgl\theta - Xl + Yl\theta,$$
or $\quad \{M(\tfrac{2}{5}a^2 + l^2) + \tfrac{1}{2}M'l^2\}\ddot{\theta} + (M - M')gl\theta = 0.$

Whence if T be the time of a small oscillation, we have
$$T = 2\pi \sqrt{\frac{\tfrac{2}{5}Ma^2 + (M + \tfrac{1}{2}M')l^2}{(M - M')gl}} \quad \ldots\ldots\ldots\ldots(41).$$

If in (41) we put $\rho = 0$, or $M' = 0$, we shall obtain the period of a pendulum which is vibrating in vacuo, and (41) shows that the effect of the liquid is to increase the period.

This result is in accordance with a general dynamical principle (to which however there are certain exceptions), *that when a dynamical system is subject to constraint, the periods of oscillation are usually greater than when the system is free.*

54. In the last example we have supposed that the liquid extends to infinity in all directions. In practice this is impossible, and it is therefore desirable to ascertain what effect the vessel which contains the liquid, produces on the period.

Let us therefore suppose that the sphere and liquid are contained within a rigid fixed spherical envelop of radius c, whose centre coincides with the equilibrium position of the sphere. The vertical pressure of the liquid upon the sphere is evidently equal to $M'g$; and the horizontal pressure will be of the form
$$\dot{V} F(x) + V^2 f(x),$$
where x is the distance of the centre of the sphere at time t from its mean position, and F and f are unknown functions. The moment of the latter about the point of suspension is
$$l\{\dot{V} F(x) + V^2 f(x)\};$$
and since in all problems relating to small oscillations, the squares and products of small quantities are neglected, it follows by Maclaurin's theorem that the moment is $l\dot{V}F(0)$, and therefore may be calculated on the supposition that the spheres are concentric. We must therefore first obtain the velocity potential

when the two spheres are concentric, and the inner sphere is set in motion with velocity V.

We have already shown that a solution of Laplace's equation is

$$\phi = (Ar + Br^{-2}) \cos \theta.$$

The boundary condition at the moving sphere is

$$\frac{d\phi}{dr} = V \cos \theta \quad \quad \quad \quad (42),$$

when $r = a$.

The boundary condition at the fixed envelop is

$$\frac{d\phi}{dr} = 0 \quad \quad \quad \quad (43),$$

when $r = c$. Substituting the above value of ϕ in (42) and (43), we obtain

$$A - 2Ba^{-3} = V, \quad A - 2Bc^{-3} = 0 \ ;$$

whence
$$A = -\frac{Va^3}{c^3 - a^3}, \quad B = -\frac{Va^3c^3}{2(c^3 - a^3)}$$

and
$$\phi = -\frac{Va^3}{c^3 - a^3}\left(r + \frac{c^3}{2r^2}\right) \cos \theta \quad \quad \quad \quad (44).$$

The pressure on the sphere may thus be obtained from our previous results by writing $\tfrac{1}{2}Va\,(c^3 + 2a^3)/(c^3 - a^3)$ for $\tfrac{1}{2}Va$; whence

$$X = \tfrac{1}{2}M'\dot{V}(c^3 + 2a^3)/(c^3 - a^3), \quad Y = M'g,$$

and the equation of motion is

$$\{M\,(\tfrac{2}{5}a^2 + l^2) + \tfrac{1}{2}M'l^2\,(c^3 + 2a^3)/(c^3 - a^3)\}\,\ddot{\theta} + (M - M')\,gl\theta = 0.$$

This equation shows that the effect of the spherical envelop, which produces an additional constraint, is to increase the period, as ought to be the case in conformity with the dynamical principle stated above; and that its action is equivalent to an increase in the inertia of the pendulum which is equal to

$$\tfrac{1}{2}M'\,(c^3 + 2a^3)/(c^3 - a^3).$$

EXAMPLES.

1. Prove that $\phi = \log \dfrac{(x+a)^2 + y^2}{(x-a)^2 + y^2}$

gives a possible motion in two dimensions. Find the form of the stream lines, and prove that the curves of equal velocity are lemniscates.

2. In the irrotational motion of a liquid, prove that the motion derived from it, by turning the direction of motion at each point in one direction through $90°$ without changing the velocity, will also be a possible irrotational motion, the conditions at the boundaries being altered so as to suit the new motion.

Discuss the motion obtained in this way from the preceding example.

3. Liquid is moving irrotationally in two dimensions, between the space bounded by the two lines $\theta = \pm \tfrac{1}{6}\pi$ and the curve $r^3 \cos 3\theta = a^3$. The bounding curves being at rest, prove that the velocity potential is of the form

$$\phi = r^3 \sin 3\theta.$$

4. The space between the elliptic cylinder $(x/a)^2 + (y/b)^2 = 1$, and a similarly situated and coaxial cylinder bounded by planes perpendicular to the axis, is filled with liquid, and made to rotate with angular velocity ω about a fixed axis. Prove that the velocity potential with reference to the principal axes of the cylinder is $\omega xy (a^2 - b^2)/(a^2 + b^2)$, and that the surfaces of equal pressure when the angular velocity is constant, are the hyperbolic cylinders

$$\frac{x^2}{3a^2 + b^2} - \frac{y^2}{3b^2 + a^2} = C.$$

5. If $\phi = f(x, y)$, $\psi = F(x, y)$ are the velocity potential and current function of a liquid, and if we write

$$x = f(\phi, \psi), \ y = F(\phi, \psi)$$

and from these expressions find ϕ and ψ; prove that the new values of ϕ and ψ will be the velocity potential and current function of some other motion of a liquid.

Hence prove that if $\phi = x^2 - y^2$, $\psi = 2xy$, the transformation gives the motion of a liquid in the space bounded by two confocal and coaxial parabolic cylinders.

6. In example 4 prove that the paths of the particles relative to the cylinder are similar ellipses, and that the paths in space are similar to the pericycloid

$$x = (a+b)\cos\theta + (a-b)\cos\left(\frac{a+b}{a-b}\right)^2\theta,$$

$$y = (a+b)\sin\theta + (a-b)\sin\left(\frac{a+b}{a-b}\right)^2\theta.$$

7. Water is enclosed in a vessel bounded by the axis of y and the hyperbola $2(x^3 - 3y^2) + x + my = 0$, and the vessel is set rotating about the axis of z. Prove that

$$\phi = 2(3x^2y - y^3) + xy - \tfrac{1}{2}m(x^2 - y^2),$$
$$\psi = 2(x^3 - 3xy^2) + \tfrac{1}{2}(x^2 - y^2) + mxy.$$

8. The space between two confocal coaxial elliptic cylinders is filled with liquid which is at rest. Prove that if the outer cylinder be moved with velocity U parallel to the major axis, and the inner with relative velocity V in the same direction, the velocity potential of the initial motion will be

$$\phi = Uc\cosh\eta\cos\xi - Vc\frac{\cosh(\beta - \eta)}{\cosh(\beta - \alpha)}\sinh\alpha\cos\xi,$$

where $\eta = \beta$, $\eta = \alpha$ are the equations of the outer and inner cylinders respectively, and $2c$ the distance between their foci.

9. If in the last example the outer cylinder were to rotate with angular velocity Ω, and the inner with angular velocity ω, prove that initially

$$\phi = \tfrac{1}{4}\Omega c^2 \frac{\cosh 2(\eta - \alpha)}{\sinh 2(\beta - \alpha)}\sin 2\xi - \tfrac{1}{4}\omega c^2 \frac{\cosh 2(\beta - \eta)}{\sinh 2(\beta - \alpha)}\sin 2\xi.$$

10. The transverse section of a uniform prismatic vessel, is of the form bounded by the two intersecting hyperbolas represented by the equations

$$\sqrt{2}(x^2 - y^2) + x^2 + y^2 = a^2, \quad \sqrt{2}(y^2 - x^2) + x^2 + y^2 = b^2.$$

If the vessel be filled with water and made to rotate with angular velocity ω about its axis, prove that the initial component velocities at any point (x, y) of the water will be

$$\frac{\omega}{a^4 + b^4}\{2y^3 - 6x^2y + \sqrt{2}(a^2 - b^2)y\}$$

$$-\frac{\omega}{a^4 + b^4}\{2x^3 - 6xy^2 + \sqrt{2}(b^2 - a^2)x\}$$

respectively.

11. In the midst of an infinite mass of liquid at rest, is a sphere of radius a, which is suddenly strained into a spheroid of small ellipticity. Find the kinetic energy due to the motion of the liquid contained between the given surface, and an imaginary concentric spherical surface of radius c; and show that if this imaginary surface were a real bounding surface which could not be deformed, the kinetic energy in this case would be to that in the former case in the ratio

$$c^5 (3a^5 + 2c^5) : 2 (c^5 - a^5)^2.$$

12. The space between two coaxial cylinders is filled with liquid, and the outer is surrounded by liquid, extending to infinity, the whole being bounded by planes perpendicular to the axis. If the inner cylinder be suddenly moved with given velocity, prove that the velocity of the outer cylinder to that of the inner, will be in the ratio

$$2b^2c^2\rho : \rho(a^2b^2 - a^2c^2 + b^4 + b^2c^2) + \sigma(a^2 - b^2)(b^2 - c^2),$$

where a and b are the external and internal radii of the outer cylinder, σ its density, c the radius of the inner cylinder and ρ the density of the liquid.

13. A solid cylinder of radius a immersed in an infinite liquid, is attached to an axis about which it can turn, whose distance from the axis of the cylinder is c, and oscillates under the action of gravity. Prove that the length of the simple equivalent pendulum is

$$\frac{\frac{1}{2}a^2 + c^2(1 + \rho/\sigma)}{c(1 - \rho/\sigma)},$$

σ and ρ being the densities of the cylinder and liquid.

14. Liquid of density ρ is contained between two confocal elliptic cylinders and two planes perpendicular to their axes. The lengths of the semi-axes of the inner and outer cylinders are $c \cosh \alpha$, $c \sinh \alpha$, $c \cosh \beta$, $c \sinh \beta$ respectively. Prove that if the outer cylinder be made to rotate about its axis with angular velocity Ω, the inner cylinder will begin to rotate with angular velocity

$$\frac{\Omega \rho \operatorname{cosech} 2(\beta - \alpha)}{\rho \coth 2(\beta - \alpha) + \frac{1}{2}\sigma \sinh 4\alpha},$$

where σ is the density of the cylinder.

15. A circular cylinder of mass M, whose centre of inertia is at a distance c from its axis, is projected in an infinite liquid under the action of gravity. Prove that the centre of inertia of the cylinder and the displaced liquid will describe a parabola, while the cylinder oscillates like a pendulum of length

$$\{(M + M') k^2 + M'c^2\}/2M'c,$$

where M' is the mass of the liquid displaced, and k is the radius of gyration of the cylinder about its axis.

16. A cylinder of radius a is surrounded by a concentric cylinder of radius b, and the intervening space is filled with liquid. The inner cylinder is moved with velocity u, and the outer with velocity v along the same straight line; prove that the velocity potential is

$$\phi = \frac{b^2 v - a^2 u}{b^2 - a^2} r \cos\theta + \frac{(v - u) a^2 b^2 \cos\theta}{(b^2 - a^2) r}.$$

17. A long cylinder of given radius is immersed in a mass of liquid bounded by a very large cylindrical envelop. If the envelop be suddenly moved in a direction perpendicular to the cylinder with velocity V, the cylinder will begin to move with velocity $\tfrac{1}{2} V$, provided the density of the cylinder be three times that of the liquid.

CHAPTER III.

MOTION OF A SINGLE SOLID IN AN INFINITE LIQUID.

55. IN the previous chapter, we obtained expressions for the velocity potential and current function in several cases in which a solid was moving in a liquid; and we also worked out several problems relating to the motion of a right circular cylinder and a sphere, by first calculating the pressure, and then by integration determining the resultant force exerted by the liquid upon the solid. This method, in the case of solids other than circular cylinders and spheres, is excessively laborious; and we shall devote the present chapter to developing a dynamical theory, which will enable us to dispense with this operation.

Since the solid and the surrounding liquid constitute a single dynamical system, the pressure exerted by the latter upon the former, is an unknown reaction arising from contiguous portions of the system. Now it is shown in treatises on Rigid Dynamics, that methods exist by means of which the motion of any dynamical system can be determined without introducing any unknown reactions. We shall therefore proceed to apply these methods to the hydrodynamical problem of a solid moving in a liquid.

It will first be desirable to call attention to the following fundamental propositions, viz.

I. *The rate of change of the component of the linear momentum, parallel to an axis, of any dynamical system, is equal to the component parallel to that axis, of the impressed forces which act upon the system.*

II. *The rate of change of the component of the angular momentum of the system about any axis, is equal to the moment of the impressed forces about that axis.*

III. *If the system is not acted upon by any dissipative forces, such as internal friction which converts energy into heat, the sum of the potential and kinetic energies of the system is constant throughout the motion.*

IV. *The work done by an impulse, is equal to half the product of the impulse into the sum of the components in the direction of the impulse, of the initial and final velocities of the point at which it is applied.*

V. *If a dynamical system be set in motion by given impulses, the work done by the impulses, is greater when the system is free, than when it is subject to constraint*[1].

56. The kinetic energy of the system which we are considering, is the sum of the kinetic energies of the moving solid and liquid. The former can be calculated by the ordinary rules of Rigid Dynamics; and we shall proceed to find an expression for the latter. But before doing this, it will be advisable to consider a certain theorem due to Green, which has important applications in various branches of physics.

---

[1] The last proposition, from the name of its discoverer, is known as Bertrand's Theorem. In order to prove the theorem in a rigorous manner, a knowledge of Abstract Dynamics would be required, but the following considerations may assist the reader.

If the dynamical system is free, the whole of the given impulse is expended in producing kinetic energy; but if the system is subject to any constraint, part of the impulse will be expended in producing impulsive reactions at the points at which the constraint is applied, and therefore the impulse available for producing kinetic energy, will be less than when the system is free.

The theorem may be easily verified by solving any simple dynamical problem. For example, let a rod AB of mass m and length $2a$ be lying on a smooth horizontal table; and let the rod be set in motion by means of an impulse F, applied at B, perpendicularly to its length. The equations of motion are

$$mv = F, \quad \tfrac{1}{3} ma^2 \omega = Fa;$$

and therefore the kinetic energy, which is equal to the work done by the impulse, is

$$2F^2/m.$$

Now let the rod be capable of rotating about the end A, which is supposed to be fixed; then the equation of motion is

$$\tfrac{4}{3} ma^2 \omega = 2Fa;$$

and the kinetic energy is

$$3F^2/2m,$$

which is less than in the former case.

Green's Theorem.

Let ϕ and ψ be any two functions, which throughout the interior of a closed surface S are single valued, and which together with their first and second derivatives, are finite and continuous at every point within S; then

$$\iiint \left(\frac{d\phi}{dx}\frac{d\psi}{dx} + \frac{d\phi}{dy}\frac{d\psi}{dy} + \frac{d\phi}{dz}\frac{d\psi}{dz} \right) dx\,dy\,dz$$

$$= \iint \phi \frac{d\psi}{dn} dS - \iiint \phi \nabla^2 \psi \, dx\,dy\,dz \quad \ldots (1)$$

$$= \iint \psi \frac{d\phi}{dn} dS - \iiint \psi \nabla^2 \phi \, dx\,dy\,dz \quad \ldots (2),$$

where the triple integrals extend throughout the volume of S, and the surface integrals over the surface of S, and dn denotes an element of the normal to S drawn outwards.

Integrating the left-hand side by parts, we obtain

$$\iiint \frac{d\phi}{dx}\frac{d\psi}{dx} dx\,dy\,dz = \left[\iint \phi \frac{d\psi}{dx} dy\,dz \right] - \iiint \phi \frac{d^2\psi}{dx^2} dx\,dy\,dz \ldots(3),$$

where the brackets denote that the double integral is to be taken between proper limits. Now since the surface is a closed surface, any line parallel to x, which enters the surface a given number of times, must issue from it the same number of times; also the x-direction cosine of the normal at the point of entrance, will be of contrary sign to the same direction cosine at the corresponding point of exit; hence the surface integral

$$= \iint \phi \frac{d\psi}{dx} l\,dS.$$

Treating each of the other terms in a similar manner, we find that the left-hand side of (3)

$$= \iint \phi \frac{d\psi}{dn} dS - \iiint \phi \nabla^2 \psi \, dx\,dy\,dz.$$

The second equation (2) is obtained by interchanging ϕ and ψ.

57. We may deduce several important corollaries.

(i) Let $\psi = 1$, and let ϕ be the velocity potential of a liquid; then $\nabla^2 \phi = 0$, and we obtain

$$0 = \iiint \nabla^2 \phi \, dx\,dy\,dz = \iint \frac{d\phi}{dn} dS \ldots\ldots\ldots\ldots(4).$$

The right-hand side is the analytical expression for the fact that the total flux across the closed surface is zero; in other words as much liquid enters the surface as issues from it.

(ii) Let ϕ and ψ be both velocity potentials, then

$$\iint \phi \frac{d\psi}{dn} dS = \iint \psi \frac{d\phi}{dn} dS \dots \dots \dots \dots (5).$$

(iii) Let $\phi = \psi$, where ϕ is the velocity potential of a liquid; then

$$\iiint \left\{ \left(\frac{d\phi}{dx}\right)^2 + \left(\frac{d\phi}{dy}\right)^2 + \left(\frac{d\phi}{dz}\right)^2 \right\} dx\,dy\,dz = \iint \phi \frac{d\phi}{dn} dS \dots (6).$$

If we multiply both sides of (6) by $\tfrac{1}{2}\rho$, the left-hand side is equal to the kinetic energy of the liquid; and the equation shows that the kinetic energy of a liquid whose motion is acyclic and irrotational, which is contained within a closed surface, depends solely upon the motion of the surface.

58. Let us now suppose that liquid contained within such a surface is originally at rest, and let the liquid be set in motion by means of an impulsive pressure p applied to every point of the surface. The motion produced must be necessarily irrotational, and acyclic; also if ϕ be its velocity potential, it follows from § 23 (34) that $p = -\rho\phi$. Now by the fourth proposition of § 55, the work done by the impulse is equal to

$$= -\tfrac{1}{2} \iint p \frac{d\phi}{dn} dS = \tfrac{1}{2}\rho \iint \phi \frac{d\phi}{dn} dS;$$

whence equation (6) shows that the work done by the impulse, is equal to the kinetic energy of the motion produced by it, as ought to be the case.

59. Let us in the next place suppose that liquid is contained within a closed surface which is in motion; and let the motion of the liquid be irrotational and acyclic; also let the surface be suddenly reduced to rest. Then if ϕ be the new velocity potential, $d\phi/dn = 0$, and therefore

$$\iiint \left\{ \left(\frac{d\phi}{dx}\right)^2 + \left(\frac{d\phi}{dy}\right)^2 + \left(\frac{d\phi}{dz}\right)^2 \right\} dx\,dy\,dz = 0,$$

whence $d\phi/dx$, $d\phi/dy$, and $d\phi/dz$ are each zero, and therefore the liquid is reduced to rest.

60. In proving Green's Theorem, we have supposed that the region through which we integrate, is contained within a single closed surface, but if the region were bounded externally and internally by two or more closed surfaces, the theorem would still be true, provided we take the surface integral with the positive sign over the external boundary, and with the negative sign over each of the internal boundaries.

61. Let us suppose that the liquid is bounded internally by one or more closed surfaces S_1, S_2 &c., and externally by a very large fixed sphere whose centre is the origin. If T be the kinetic energy of the liquid,

$$T = \tfrac{1}{2}\rho \iint \phi \frac{d\phi}{dn} dS - \tfrac{1}{2}\rho \left[\iint \phi \frac{d\phi}{dn} dS \right],$$

where the square brackets indicate that the integral is to be taken over each of the internal boundaries.

If the liquid be at rest at infinity, the value of ϕ at S, will be at most of the order m/r, where m is a constant, and

$$d\phi/dn = d\phi/dr = - m/r^2;$$

also if $d\Omega$ be the solid angle subtended by dS at the origin, $dS = r^2 d\Omega$; therefore

$$\iint \phi \frac{d\phi}{dn} dS = - \frac{m^2}{r} \iint d\Omega = - \frac{4\pi m^2}{r},$$

which vanishes when $r = \infty$. Hence the kinetic energy of an infinite liquid bounded internally by closed surfaces is

$$T = - \tfrac{1}{2}\rho \left[\iint \phi \frac{d\phi}{dn} dS \right] \quad \ldots\ldots\ldots\ldots\ldots\ldots\ldots(7),$$

where the surface integral is to be taken over each of the internal boundaries.

The preceding expression for the kinetic energy shows, that if the motion is acyclic and the *internal* boundaries of the liquid be suddenly reduced to rest, the whole liquid will be reduced to rest.

62. When a single solid is moving in an infinite liquid, the velocity potential must satisfy the following conditions;

(i) ϕ must be a single valued function, which at all points of the liquid satisfies the equation $\nabla^2 \phi = 0$,

(ii) ϕ and its first derivatives must be finite and continuous at all points of the liquid, and must vanish at infinity, if any portion of the liquid extends to infinity,

(iii) At all points of the liquid which are in contact with a moving solid, $d\phi/dn$ must be equal to the normal velocity of the solid, where dn is an element of the normal to the solid drawn outwards; if any portion of the liquid is in contact with fixed boundaries, $d\phi/dn$ must be zero at every point of these fixed boundaries.

The most general possible motion of a solid, may be resolved into three component velocities parallel to three rectangular axes, (which may either be fixed or in motion), together with three angular velocities about these axes.

Let us therefore refer the motion to three rectangular axes Ox, Oy, Oz *fixed* in the solid, and let ϕ_1 be the velocity potential when the solid is moving with unit velocity parallel to Ox, and let χ_1 be the velocity potential when the solid is rotating with unit angular velocity about Ox. Let ϕ_2, ϕ_3, χ_2, χ_3 be similar quantities with respect to Oy and Oz. Also let u, v, w be the linear velocities of the solid parallel to, and ω_1, ω_2, ω_3 be its angular velocities about the axes.

We can now show that the velocity potential of the whole motion will be

$$\phi = u\phi_1 + v\phi_2 + w\phi_3 + \omega_1\chi_1 + \omega_2\chi_2 + \omega_3\chi_3 \ldots\ldots\ldots(8).$$

For if λ, μ, ν be the direction cosines of the normal at any point x, y, z on the surface of the solid, we must have at the surface

$$\frac{d\phi_1}{dn} = \lambda, \quad \frac{d\phi_2}{dn} = \mu, \quad \frac{d\phi_3}{dn} = \nu,$$

$$\frac{d\chi_1}{dn} = \nu y - \mu z, \quad \frac{d\chi_2}{dn} = \lambda z - \nu x, \quad \frac{d\chi_3}{dn} = \mu x - \lambda y.$$

Hence $\dfrac{d\phi}{dn} = (u - y\omega_3 + z\omega_2)\lambda + (v - z\omega_1 + x\omega_3)\mu + (w - x\omega_2 + y\omega_1)\nu$

= normal velocity of the solid.

63. If we substitute the value of ϕ from (8) in (7), it follows that T is a homogeneous quadratic function of the six velocities u, v, w, ω_1, ω_2, ω_3, and therefore contains twenty-one terms. If we choose as our axes Ox, Oy, Oz, the principal axes at the centre

KINETIC ENERGY. 65

of inertia O of the solid, the kinetic energy of the latter will be equal to

$$\tfrac{1}{2}M(u^2+v^2+w^2)+\tfrac{1}{2}(A_1\omega_1^2+B_1\omega_2^2+C_1\omega_3^2)$$

where M is the mass of the solid, and A_1, B_1, C_1 are its principal moments of inertia. Hence the kinetic energy T of the system, being the sum of the kinetic energies of the solid and liquid, is determined by the equation,

$$\begin{aligned}2T = {} & Pu^2+Qv^2+Rw^2+2P'vw+2Q'wu+2R'uv\\
& +A\omega_1^2+B\omega_2^2+C\omega_3^2+2A'\omega_2\omega_3+2B'\omega_3\omega_1+2C'\omega_1\omega_2\\
& +2\omega_1(Lu+Mv+Nw)\\
& +2\omega_2(L'u+M'v+N'w)\\
& +2\omega_3(L''u+M''v+N''w)\dots\dots\dots\dots\dots\dots(9).\end{aligned}$$

The coefficients of the velocities in the preceding expression for the kinetic energy, are called *coefficients of inertia*. The quantities P, Q, R are called the *effective inertias of the solid parallel to the principal axes*, and the quantities A, B, C are called the *effective moments of inertia about the principal axes*. If the liquid extend to infinity, and there is only one moving solid, the coefficients of inertia depend solely upon the form and density of the solid and the density of the liquid, and not upon the coordinates which determine the position of the solid in space.

The values of these coefficients are

$$\left.\begin{aligned}P &= M-\rho\iint\phi_1\frac{d\phi_1}{dn}dS = M-\rho\iint\phi_1\lambda dS\\
P' &= -\tfrac{1}{2}\rho\iint\phi_2\frac{d\phi_3}{dn}dS-\tfrac{1}{2}\rho\iint\phi_3\frac{d\phi_2}{dn}dS = -\rho\iint\phi_2\frac{d\phi_3}{dn}dS\end{aligned}\right\}\dots(10)$$

by (5); with similar expressions for the other coefficients.

When the form of the solid resembles that of an ellipsoid, which is symmetrical with respect to three perpendicular planes through its centre of inertia, and the motion is referred to the principal axes of the solid at that point, the kinetic energy must remain unchanged when the direction of any one of the component velocities is reversed; hence the kinetic energy cannot contain any of the products of the velocities, and must therefore be of the form;

$$2T = Pu^2+Qv^2+Rw^2+A\omega_1^2+B\omega_2^2+C\omega_3^2\dots\dots(11).$$

If in addition, the solid is one of revolution about the axis of z, the kinetic energy will not be altered if u is changed into v, and ω_1 into ω_2; whence $P = Q$, $A = B$, and

$$2T = P(u^2 + v^2) + Rw^2 + A(\omega_1^2 + \omega_2^2) + C\omega_3^2 \ldots\ldots(12).$$

Although every solid of revolution must be symmetrical with respect to all planes through its axis, it is not necessarily symmetrical with respect to a plane perpendicular to its axis. The solid formed by the revolution of a cardioid about its axis is an example of such a solid. In this case the kinetic energy will be unaltered when the signs of u, v or ω_3 are changed, and also when u is changed into v, and ω_1 into ω_2; hence in this case

$$2T = P(u^2 + v^2) + Rw^2 + A(\omega_1^2 + \omega_2^2) + C\omega_3^2 + 2Nw(\omega_1 + \omega_2)\ldots(13).$$

If the solid moves with its axis in one plane, (say zx), v and ω_1 must be zero, and the last term may be got rid of, by moving the origin to a point on the axis of z, whose distance from the origin is $-N/R$. This point is called the *Centre of Reaction*.

64. We must now find expressions for the component linear and angular momenta.

Since we are confining our attention to acyclic irrotational motion, it follows from § 57 that the motion of the liquid at any instant, depends solely upon the motion of the surface of the solid; hence the motion which actually exists at any particular epoch, could be produced instantaneously from rest, by the application of suitable impulsive forces to the solid; and since impulsive forces are measured by the momenta which they produce, it follows that the resultant impulse which must be applied to the solid, must be equal to resultant momentum of the solid and liquid.

Let ξ, η, ζ be the components parallel to the principal axes of the solid, of the impulsive force which must be applied to the solid, in order to produce the actual motion which exists at time t; and let λ, μ, ν be the components about these axes, of the impulsive couple. Then ξ, η, ζ are the components of the linear momentum, and λ, μ, ν of the angular momentum of the solid and liquid.

Let p denote the impulsive pressure of the liquid, and let us consider the effect produced upon the solid by the application of the impulse whose components are ξ, η, ζ, λ, μ, ν.

By the ordinary equations of impulsive motion

$$Mu = \xi - \iint pldS,$$

where l, m, n are the direction cosines of the normal at any point of S.

But if ϕ be the velocity potential of the motion instantaneously generated by the impulse, and which is equal to the velocity potential which actually exists at time t, it follows from § 23, that $p = -\rho\phi$, whence

$$\xi = Mu - \rho\iint \phi l \, dS$$

$$= Mu - \rho \iint \phi \frac{d\phi_1}{dn} \, dS,$$

since at the surface of the solid

$$l = d\phi_1/dn.$$

If $\mathbb{T}'$, $\mathbb{T}$ be the kinetic energies of the solid and liquid, it follows from (7) and (8) that

$$\frac{d\mathbb{T}}{du} = -\tfrac{1}{2}\rho \iint \phi \frac{d\phi_1}{dn} \, dS - \tfrac{1}{2}\rho \iint \phi_1 \frac{d\phi}{dn} \, dS$$

$$= -\rho \iint \phi \frac{d\phi_1}{dn} \, dS,$$

since by Green's Theorem both the double integrals are equal. Also

$$\frac{d\mathbb{T}'}{du} = Mu,$$

whence
$$\xi = \frac{d\mathbb{T}'}{du} + \frac{d\mathbb{T}}{du}$$

$$= \frac{dT}{du}.$$

We therefore see that the component momentum along the axis of x, is equal to the differential coefficient of the kinetic energy, with respect to the component velocity of the solid along the same axis; and by precisely similar reasoning, it can be shown that the component angular momentum about the axis of x, is equal to the differential coefficient of the kinetic energy, with respect to the component angular velocity of the solid about this axis. We thus obtain the following equations for determining the momenta, viz.:

$$\left. \begin{aligned} \xi = \frac{dT}{du}, \quad \eta = \frac{dT}{dv}, \quad \zeta = \frac{dT}{dw} \\ \lambda = \frac{dT}{d\omega_1}, \quad \mu = \frac{dT}{d\omega_2}, \quad \nu = \frac{dT}{d\omega_3} \end{aligned} \right\} \ldots\ldots\ldots\ldots(14).$$

Equations (14) are well-known dynamical equations.

65. The preceding expressions for the kinetic energy and momenta, have been obtained in a direct manner, by means of hydrodynamical principles; but the reader who desires a short cut to equations (9) and (14), may begin by assuming the theorem, that the kinetic energy of a dynamical system is a homogeneous quadratic function of the velocities of the system. Since the kinetic energy of a liquid which surrounds a single moving solid, and whose motion is acyclic and irrotational, must vanish when the solid is reduced to rest, the kinetic energy of the liquid must be a homogeneous quadratic function of the velocities of the *solid*. This leads to (9). Also since the kinetic energy of an *infinite* liquid, when expressed in the form (9), cannot depend upon the position of the solid in space, the coefficients of the velocities must be constant quantities.

If ξ, η, ζ, λ, μ, ν be the component impulses, which must be applied to the solid, in order to generate from rest, the motion which actually exists at time t, it follows from the fourth proposition of § 55, that

$$2T = \xi u + \eta v + \zeta w + \lambda \omega_1 + \mu \omega_2 + \nu \omega_3.$$

But since T is a homogeneous quadratic function of the velocities,

$$2T = u\frac{dT}{du} + v\frac{dT}{dv} + w\frac{dT}{dw} + \omega_1\frac{dT}{d\omega_1} + \omega_2\frac{dT}{d\omega_2} + \omega_3\frac{dT}{d\omega_3}.$$

Comparing these two equations, we obtain (14).

We are now in a position to solve a variety of problems connected with the motion of a single solid in an infinite liquid.

Motion of a Sphere.

66. Let us suppose that the centre of the sphere describes a plane curve, and let u, v be its component velocities parallel to the axes of x and y. Since every diameter of a sphere is a principal axis, the axes of x and y may be supposed to be fixed in direction, whence

$$\phi = -\frac{a^3}{2r^3}(ux + vy);$$

and since on account of symmetry $P = Q$, we have

$$T = \tfrac{1}{2}P(u^2 + v^2),$$

where
$$P = M - \rho \iint \phi_1 l \, dS$$
$$= M + \pi \rho a^3 \int_0^\pi \cos^2\theta \sin\theta \, d\theta$$
$$= M + \tfrac{1}{2} M',$$
where M' is the mass of the liquid displaced. Whence
$$T = \tfrac{1}{2}(M + \tfrac{1}{2}M')(u^2 + v^2),$$
and therefore
$$\xi = (M + \tfrac{1}{2}M')u, \quad \eta = (M + \tfrac{1}{2}M')v.$$

Let us now suppose that the sphere is descending under the action of gravity, and that the axis of y is drawn vertically downwards; we shall also suppose that the sphere is initially projected with a velocity, whose horizontal and vertical components are U and V.

Since the momentum parallel to x is constant throughout the motion,
$$(M + \tfrac{1}{2}M')u = \text{const.} = (M + \tfrac{1}{2}M')U,$$
whence
$$u = U.$$

The equation giving the vertical motion is,
$$\frac{d\eta}{dt} = (M - M')g,$$
or
$$(M + \tfrac{1}{2}M')\frac{dv}{dt} = (M - M')g.$$

Whence if σ be the density of the sphere,
$$\frac{dv}{dt} = \frac{\sigma - \rho}{\sigma + \tfrac{1}{2}\rho} g,$$
and the sphere will describe a parabola with vertical acceleration $g(\sigma - \rho)/(\sigma + \tfrac{1}{2}\rho)$, in accordance with our previous result.

The motion of a right circular cylinder can be investigated in a precisely similar manner.

Motion of an Elliptic Cylinder.

67. Let u, v be the velocities of the cylinder parallel to the major and minor axes of its cross section, ω its angular velocity about its axis. On account of symmetry none of the products can appear, and therefore
$$T = \tfrac{1}{2}(Pu^2 + Qv^2 + A\omega^2) \quad \ldots\ldots\ldots\ldots\ldots\ldots (15),$$
where P, Q, A are constant quantities.

Let us now suppose that no forces are in action, and that the solid and liquid are initially at rest; and let the cylinder be set in motion by means of a linear impulse F, whose line of action passes through its axis, and a couple which produces an initial angular velocity Ω.

Let us refer the motion to two fixed rectangular axes x and y, the former of which coincides with the direction of F, and let θ be the angle which the major axis of the cross section makes with the axis of x at time t.

Resolving the momenta along the axes of x and y, we obtain

$$\xi \cos \theta - \eta \sin \theta = F$$
$$\xi \sin \theta + \eta \cos \theta = 0,$$

whence since

$$\xi = Pu, \quad \eta = Qv,$$

we obtain

$$Pu = F \cos \theta, \quad Qv = -F \sin \theta \ldots\ldots\ldots\ldots(16).$$

Since the kinetic energy remains constant throughout the motion, it follows that if we substitute the values of u, v from (16) in (15), and put β for the initial value of θ, we shall obtain

$$F^2 \left(\frac{\cos^2 \theta}{P} + \frac{\sin^2 \theta}{Q} \right) + A \dot\theta^2 = F^2 \left(\frac{\cos^2 \beta}{P} + \frac{\sin^2 \beta}{Q} \right) + A \Omega^2,$$

or

$$A \dot\theta^2 = A \Omega^2 + F^2 \left(\frac{1}{P} - \frac{1}{Q} \right) (\sin^2 \theta - \sin^2 \beta) \ldots\ldots(17).$$

We shall presently show that $Q > P$; it therefore follows that if

$$\Omega < F \sin \beta \sqrt{\frac{Q-P}{APQ}},$$

$\dot\theta$ will vanish, and the cylinder will oscillate; but if

$$\Omega > F \sin \beta \sqrt{\frac{Q-P}{APQ}},$$

$\dot\theta$ will never vanish, and the cylinder will make a complete revolution.

The integration of equation (17) requires elliptic functions, but without introducing those quantities, we can easily ascertain the character of the motion of the centre of inertia of the cylinder.

Let (x, y) be the coordinates of the centre of inertia referred to the fixed axes of x and y; then

$$\dot x = u \cos \theta - v \sin \theta, \quad \dot y = u \sin \theta + v \cos \theta \ldots\ldots(18).$$

Substituting the values of u, v from (16) we obtain

$$\dot{x} = \frac{F}{Q} + F\left(\frac{1}{P} - \frac{1}{Q}\right)\cos^2\theta,$$

$$\dot{y} = F\left(\frac{1}{P} - \frac{1}{Q}\right)\sin\theta\cos\theta.$$

These equations show that the centre of inertia of the cross section of the cylinder, moves along a straight line parallel to the direction of F with a uniform velocity F/Q, superimposed upon which is a variable periodic velocity, and that at the same time it vibrates perpendicularly to this line. This kind of motion frequently occurs in hydrodynamics, and a body moving in such a manner is called a *Quadrantal Pendulum*.

If
$$A\Omega^2 = F^2\left(\frac{1}{P} - \frac{1}{Q}\right)\sin^2\beta,$$

which is the limiting case between oscillation and rotation, the equations of motion admit of complete integration. Putting

$$I^2 = \frac{F^2}{A}\left(\frac{1}{P} - \frac{1}{Q}\right),$$

(17) becomes $\quad \dot{\theta} = I\sin\theta$

whence $\quad It = \log\tan\tfrac{1}{2}\theta.$

Therefore $\quad \dfrac{dy}{d\theta} = \dfrac{IA}{F}\cos\theta,$

$$y = \frac{IA}{F}\sin\theta,$$

$$\frac{dx}{d\theta} = \frac{F}{PI}\csc\theta - \frac{IA}{F}\sin\theta,$$

$$x = \frac{F}{PI}\log\tan\tfrac{1}{2}\theta + \frac{IA}{F}\cos\theta.$$

Putting $IA/F = c$, and eliminating θ we obtain the equation of the path, viz.

$$x = \frac{F}{PI}\log\frac{y}{c + \sqrt{c^2 - y^2}} + \sqrt{c^2 - y^2}.$$

The curves described by the centre of inertia of the cylinder in the three cases have been traced by Prof. Greenhill, and are shown in Figures 1, 2, 3 of the accompanying diagram.

72   MOTION OF A SINGLE SOLID.

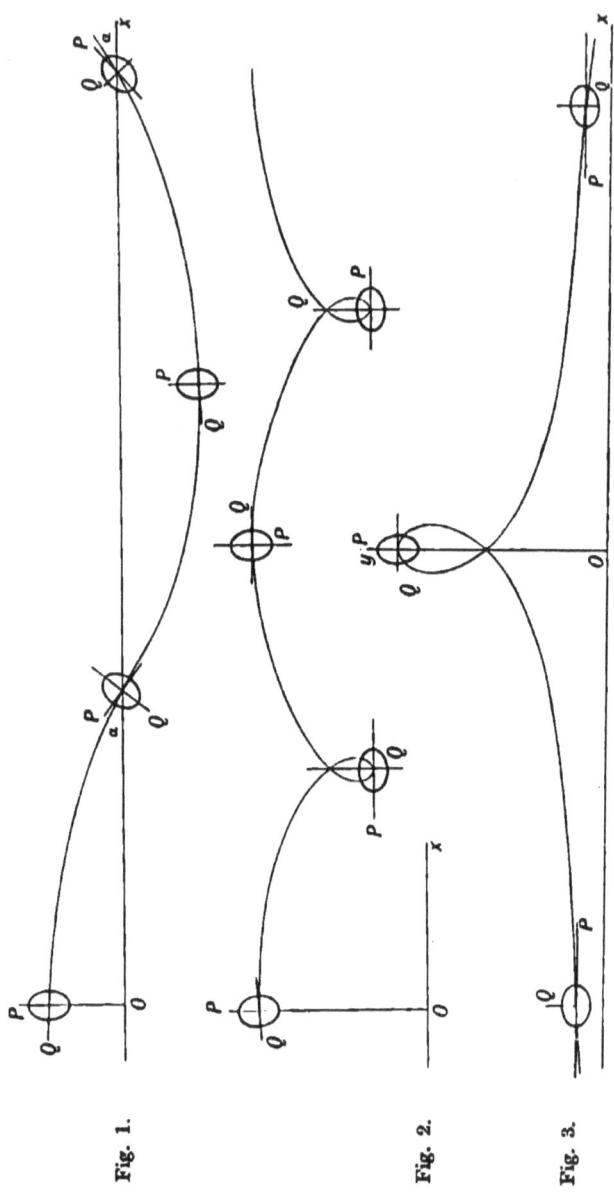

Fig. 1.

Fig. 2.

Fig. 3.

68. We shall now show that for an elliptic cylinder $Q > P$.

When the cylinder is moving with unit velocity parallel to x, we have shown in § 47 that

$$\psi_2 = c\epsilon^{-\eta+\beta} \sinh \beta \sin \xi,$$

and therefore $\quad \phi_1 = - c\epsilon^{-\eta+\beta} \sinh \beta \cos \xi.$

Now at the surface

$$-\int \phi_1 l ds = c \sinh \beta \int \cos \xi dy$$

$$= c^2 \sinh^2 \beta \int_0^{2\pi} \cos^2 \xi d\xi$$

$$= \pi b^2,$$

whence $\quad P = M - \rho \int \phi_1 l ds$

$$= M \left(1 + \frac{\rho b}{\sigma a}\right),$$

where σ is the density of the cylinder.

The value of Q is evidently obtained by interchanging a and b, whence $Q > P$.

Similar results can be proved to be true in the case of an ellipsoid; from which it is inferred *that when any solid body is moving in an infinite liquid, the effective inertias corresponding to the greatest, mean, and least principal axes, are in descending order of magnitude.*

69. If the cylinder be projected parallel to a principal axis without rotation, it will continue to move in a straight line with uniform velocity; but if the direction of projection is not a principal axis, it will begin to rotate, and its angular velocity at any subsequent time will be determined by putting $\Omega = 0$ in (17). We shall now show that if the cylinder be projected parallel to a principal axis, its motion will be stable or unstable according as the direction of projection coincides with the minor or major axis.

Let us first suppose the cylinder projected parallel to its major axis, and that a slight disturbance is communicated to it. The equation determining the angular velocity is obtained by putting $\Omega = \beta = 0$ in (17); whence

$$A \dot{\theta}^2 = F^2 \left(\frac{1}{P} - \frac{1}{Q}\right) \sin^2 \theta,$$

and therefore differentiating, and remembering that in the beginning of the disturbed motion θ is a small quantity, we obtain

$$A\ddot{\theta} + F^2\left(\frac{1}{Q} - \frac{1}{P}\right)\theta = 0.$$

Since $Q > P$, the coefficient of θ is negative, which shows that the motion is unstable.

If the cylinder is projected parallel to its minor axis we must put $\beta = \frac{1}{2}\pi$; also if $\chi = \frac{1}{2}\pi - \theta$, χ will be a small quantity in the beginning of the disturbed motion; whence (17) becomes

$$A\dot{\chi}^2 = -F^2\left(\frac{1}{P} - \frac{1}{Q}\right)\sin^2\chi,$$

whence

$$A\ddot{\chi} + F^2\left(\frac{1}{P} - \frac{1}{Q}\right)\chi = 0.$$

Since the coefficient of χ is positive, the motion is stable.

It can also be shown that if an ellipsoid be projected parallel to a principal axis, without rotation, the motion will be unstable unless the direction of projection coincides with the *least axis*. We shall however presently show, that if an ovary ellipsoid be projected parallel to its axis, the motion will be stable, provided a sufficiently large angular velocity be communicated to the solid about its axis.

70. Let us now investigate the motion of an elliptic cylinder, which descends from rest under the action of gravity.

Let the axis of y be horizontal, and the axis of x be drawn vertically downwards.

The equations of momentum are

$$\xi \sin\theta + \eta \cos\theta = 0,$$

$$\frac{d}{dt}(\xi \cos\theta - \eta \sin\theta) = (M - M')g,$$

from the last of which we obtain

$$\xi \cos\theta - \eta \sin\theta = (M - M')gt.$$

Solving these equations and recollecting that $\xi = Pu$, $\eta = Qv$, we obtain

$$Pu = (M - M')gt\cos\theta, \quad Qv = -(M - M')gt\sin\theta.$$

HELICOIDAL STEADY MOTION. 75

Substituting these values of u and v in (18), we obtain

$$\left.\begin{aligned}\dot{x} &= \left(\frac{\cos^2\theta}{P} + \frac{\sin^2\theta}{Q}\right)(M - M')gt \\ \dot{y} &= \left(\frac{1}{P} - \frac{1}{Q}\right)gt\sin\theta\cos\theta\end{aligned}\right\} \dots\dots\dots (19).$$

The equation of energy gives

$$\left(\frac{\cos^2\theta}{P} + \frac{\sin^2\theta}{Q}\right)(M - M')^2 g^2 t^2 + A\dot{\theta}^2 = 2(M - M')gx.$$

If we differentiate this equation with respect to t, we can eliminate $\dot{x}$ by means of (19); but the resulting equation would be difficult to deal with. We see however from the first of (19), that $\dot{x}$ is always positive, and therefore the cylinder moves downwards with a variable velocity, which depends upon the inclination of its major axis to the vertical, as well as upon the time. We also see from the second equation, that the horizontal velocity vanishes, whenever the major axis becomes horizontal or vertical; but if the motion should be of such a character that θ always lies between 0 and $\frac{1}{2}\pi$, the horizontal velocity will never vanish.

Helicoidal Steady Motion of a Solid of Revolution.

71. In the figure let OC be the axis of the solid of revolution, O its centre of inertia, and let the solid be rotating with angular velocity Ω about its axis.

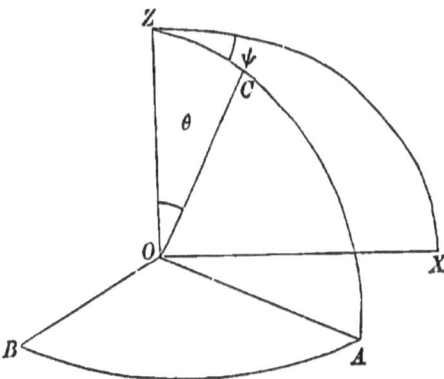

Let the solid be set in motion by means of an impulsive force F along OZ, and an impulsive couple G about OZ, and let α be the angle which OC initially makes with OZ.

Let ψ be the angle which the plane ZOC makes at time t with a plane ZOX, which is parallel to some fixed plane, and let the former plane cut the equatorial plane in OA; also let $ZOC = \theta$.

It will be convenient to refer the motion to three moving axes, OA, OB, OC, where OB is the equatorial axis which is perpendicular to OA.

Resolving the linear momentum of the system along OZ, OX and a line OY perpendicular to the plane ZOX, we obtain

$$-\xi \sin \theta + \zeta \cos \theta = F,$$
$$(\xi \cos \theta + \zeta \sin \theta) \cos \psi - \eta \sin \psi = 0,$$
$$(\xi \cos \theta + \zeta \sin \theta) \sin \psi + \eta \cos \psi = 0,$$

whence $\qquad \xi = -F \sin \theta, \quad \eta = 0, \quad \zeta = F \cos \theta \ldots\ldots\ldots\ldots (20).$

Since the components of momentum parallel to the axes of X and Y (which are fixed in direction, but not in position because O is in motion) are zero throughout the motion, the angular momentum about OZ is constant, whence

$$-A\omega_1 \sin \theta + C\Omega \cos \theta = G + C\Omega \cos \alpha \ldots\ldots\ldots (21).$$

The equation of energy gives

$$Pu^2 + Rw^2 + A(\omega_1^2 + \dot{\theta}^2) = \text{const.} \ldots\ldots\ldots\ldots (22),$$

and $\qquad \xi = Pu, \quad \zeta = Rw,$

and therefore by (20) and (21) this becomes

$$F^2 \left(\frac{\sin^2 \theta}{P} + \frac{\cos^2 \theta}{R}\right) + \frac{\{G + C\Omega(\cos \alpha - \cos \theta)\}^2}{A \sin^2 \theta} + A\dot{\theta}^2 = \text{const.}$$
$$= \text{its initial value} \ldots\ldots (23).$$

This equation determines the inclination θ of the axis.

So far our equations have been perfectly general, we shall now introduce the conditions of steady motion. These are

$$\theta = \alpha, \quad \dot{\psi} = \mu, \quad \ddot{\theta} = \dot{\theta} = 0 \ldots\ldots\ldots\ldots\ldots (24).$$

Now $\omega_1 = -\dot{\psi} \sin \alpha = -\mu \sin \alpha$, whence (21) becomes

$$A\mu \sin^2 \alpha = G \ldots\ldots\ldots\ldots\ldots\ldots\ldots (25).$$

Differentiating (23) with respect to t and using (24) and (25) we obtain

$$A\mu^2 \cos \alpha - C\Omega\mu + \left(\frac{1}{R} - \frac{1}{P}\right) F^2 \cos \alpha = 0 \ldots\ldots\ldots (26).$$

This is a quadratic equation for determining μ, when Ω and F

are given. Now μ must necessarily be a real quantity, and therefore the condition that steady motion may be possible is that

$$C^2\Omega^2 > 4F^2 A \cos^2\alpha \left(\frac{1}{R} - \frac{1}{P}\right) \quad \ldots\ldots\ldots\ldots (27),$$

and since $\quad Rw = \zeta = F \cos\alpha,$

the condition becomes

$$C^2\Omega^2 > 4AR^2w^2 \left(\frac{1}{R} - \frac{1}{P}\right) \quad \ldots\ldots\ldots\ldots (28).$$

If the solid of revolution is oblate (such as a planetary ellipsoid) $R > P$, and therefore (28) is always satisfied; but if the solid is prolate (such as an ovary ellipsoid) $P > R$, and therefore steady motion will not be possible unless (28) is satisfied.

In order to find the path described by the centre of inertia of the solid in steady motion, we have, since $\psi = \mu t$,

$$\dot{x} = (u\cos\alpha + w\sin\alpha)\cos\psi = F\left(\frac{1}{R} - \frac{1}{P}\right)\sin\alpha\cos\alpha\cos\mu t,$$

$$\dot{y} = (u\cos\alpha + w\sin\alpha)\sin\psi = F\left(\frac{1}{R} - \frac{1}{P}\right)\sin\alpha\cos\alpha\sin\mu t,$$

$$\dot{z} = w\cos\alpha - u\sin\alpha \quad\quad = F\left(\frac{\sin^2\alpha}{P} + \frac{\cos^2\alpha}{R}\right),$$

which shows that the centre of inertia describes a helix.

72. In order to find whether the steady motion is stable or unstable, differentiate (23) with respect to t, and we obtain

$$A\ddot{\theta} + f(\theta) = 0 \ldots\ldots\ldots\ldots\ldots\ldots\ldots(29),$$

where

$$f(\theta) = \tfrac{1}{2}F^2\left(\frac{1}{P} - \frac{1}{R}\right)\sin 2\theta + \frac{C\Omega}{A\sin\theta}\{G + C\Omega(\cos\alpha - \cos\theta)\}$$

$$- \frac{\cos\theta}{A\sin^2\theta}\{G + C\Omega(\cos\alpha - \cos\theta)\}^2.$$

The condition for steady motion is, that $f(\alpha) = 0$, which leads to (26), whence writing $\theta = \alpha + \chi$ where χ is small, (29) becomes

$$A\ddot{\chi} + f'(\alpha)\chi = 0,$$

and the condition of stability requires that $f'(\alpha)$ should be positive. Now

$$f'(\alpha) = A\mu^2(1 + 2\cos^2\alpha) - 3C\Omega\mu\cos\alpha + \frac{C^2\Omega^2}{A} - F^2\left(\frac{1}{R} - \frac{1}{P}\right)\cos 2\alpha,$$

whence eliminating Ω by (26) this becomes

$$A^2\mu^2 f'(\alpha) = A^2\mu^4 + A\mu^2 F^2\left(\frac{1}{R} - \frac{1}{P}\right)(1 - 3\cos^2\alpha) + F^4\left(\frac{1}{R} - \frac{1}{P}\right)^2 \cos^2\alpha$$

The condition that the right-hand side should be positive is that

$$A^2 F^4\left(\frac{1}{R} - \frac{1}{P}\right)^2 \sin^2\alpha\,(9\cos^2\alpha - 1) > 0,$$

which requires that α should lie between $\cos^{-1}\frac{1}{3}$ and 0, or between $\pi - \cos^{-1}\frac{1}{3}$ and π.

As a particular example let the solid be projected point foremost; then $\alpha = 0$ and $G = 0$, and therefore since θ is a small quantity in the beginning of the disturbed motion

$$f(\theta) = \left\{\frac{C^2\Omega^2}{4A} + F^2\left(\frac{1}{P} - \frac{1}{R}\right)\right\}\theta.$$

If therefore $R > P$ the motion is always stable, whether there is or is not rotation, and consequently the forward motion of a planetary ellipsoid is always stable; but if $P > R$, it follows that since $F = Rw$, the motion will be unstable unless

$$\Omega > \frac{2Rw}{C}\sqrt{A\left(\frac{1}{R} - \frac{1}{P}\right)}.$$

The motion of an ovary ellipsoid is therefore unstable, unless the ratio of its angular velocity to its forward velocity exceeds a certain value.

These results have an application in gunnery.

When an elongated body, such as a bullet, is fired from a gun with a high velocity, the effect of the air upon its motion cannot be neglected; and if the air is treated as an incompressible fluid, the previous investigation shows that the bullet will tend to present its flat side to the air, and also to deviate from its approximately parabolic path, unless it be endowed with a rapid rotation about its axis. Hence the bores of all guns destined for long ranges are rifled, by means of which a rapid rotation is communicated to the bullet before it leaves the barrel. The effect of the rifling tends to keep the bullet moving point foremost, and to ensure its travelling along an approximately parabolic path in a vertical plane. Moreover when a bullet is moving with a high velocity, the effect of friction cannot be entirely neglected; and it is tolerably obvious that when the bullet is moving with its flat side foremost, the effect of frictional resistance will be much greater than when it is moving point foremost, and therefore the bullet will not carry so far in the former as in the latter case. The hydrodynamical theory therefore explains the necessity of rifling guns.

Motion of a Cylinder parallel to a Plane.

73. We have thus far supposed that the liquid extends to infinity in all directions; we shall now suppose that the liquid is bounded by a fixed plane, and shall enquire what effect the plane boundary produces on the motion of a circular cylinder.

Let the axis of y be drawn perpendicularly to the plane, and let the origin be in the plane, and let (x, y) be the coordinates of the centre of the cylinder, (u, v) its velocities parallel to and perpendicular to the plane.

The kinetic energy of the solid and liquid must be a homogeneous quadratic function of u and v, but since the kinetic energy is necessarily unchanged when the sign of u is reversed, the product uv cannot appear. We may therefore write

$$T = \tfrac{1}{2}(Ru^2 + R'v^2) \ldots\ldots\ldots\ldots\ldots\ldots(30).$$

The coefficients R, R' depend upon the distance of the cylinder from the plane, and are therefore functions of y but not of x; and as a matter of fact their values are equal. It will not however be necessary to assume the equality of R and R', since our object will be attained provided we can show that R and R' diminish as y increases.

In order to produce from rest the motion which actually exists at time t, we must apply to the cylinder impulsive forces whose components are X, Y; and we must also apply at every point of the plane boundary an impulsive pressure, which is just sufficient to prevent the liquid in contact with the plane, from having any velocity perpendicular to the plane. The work done by the impulsive pressure is zero, whilst the work done by the impulses X, Y is

$$\tfrac{1}{2}(Xu + Yv) \ldots\ldots\ldots\ldots\ldots(31),$$

which must be equal to T. Now (30) may be written in the form

$$T = \tfrac{1}{2}\left(u\frac{dT}{du} + v\frac{dT}{dv}\right) \ldots\ldots\ldots\ldots(32),$$

whence comparing (31) and (32) we see that

$$X = \frac{dT}{du} = Ru, \quad Y = \frac{dT}{dv} = R'v \quad \ldots\ldots\ldots (33),$$

and therefore

$$T = \tfrac{1}{2}\left(\frac{X^2}{R} + \frac{Y^2}{R'}\right) \quad \ldots\ldots\ldots\ldots\ldots (34).$$

The last equation gives the kinetic energy produced by *given* impulses X, Y.

Let us now suppose that the cylinder instead of being at a distance y from the plane, is at a distance y_1, where $y_1 > y$; and let R_1, R_1' be the values of R, R' at y_1. Then if the cylinder were set in motion by the same impulses, the work done would be

$$T_1 = \tfrac{1}{2}\left(\frac{X^2}{R_1} + \frac{Y^2}{R_1'}\right) \quad \ldots\ldots\ldots\ldots\ldots (35).$$

Now the effect of the plane boundary is to produce a constraint, and the effect of this constraint evidently diminishes as the distance of the cylinder from the plane increases, and therefore by the fifth proposition stated at the commencement of this Chapter, $T_1 > T$. Hence

$$X^2\left(\frac{1}{R_1} - \frac{1}{R}\right) + Y^2\left(\frac{1}{R_1'} - \frac{1}{R'}\right)$$

is positive for all values of X and Y, which requires that

$$R > R_1, \quad R' > R_1';$$

hence R, R' diminish as y increases, and consequently their differential coefficients with respect to y are *negative*.

74. We can now determine the motion of the cylinder.

The momentum parallel to x is equal to dT/du and is constant; whence

$$Ru = \text{const.} = G \quad \ldots\ldots\ldots\ldots\ldots (36).$$

Since the kinetic energy is constant, we have

$$Ru^2 + R'v^2 = \text{const.} = 2T \ldots\ldots\ldots\ldots (37).$$

Differentiating (37) with respect to t, and eliminating du/dt by (36), we obtain

$$\dot{v} = -\frac{1}{2R'}\left(v^2\frac{dR'}{dy} - u^2\frac{dR}{dy}\right) \quad \ldots\ldots\ldots\ldots (38).$$

From this equation we can ascertain the effect of the plane

EFFECT OF A PLANE BOUNDARY. 81

boundary; for if the cylinder is projected perpendicularly to the plane, $u = 0$, and

$$\dot{v} = -\frac{v^2}{2R'}\frac{dR'}{dy}.$$

Now dR'/dy is negative, and therefore $\dot{v}$ is positive; whence it follows that whether the cylinder be moving from or towards the plane, the force exerted by the liquid upon the cylinder, will always be a *repulsion* from the plane, which is equal to

$$-\frac{Mv^2}{2R'}\frac{dR'}{dy}.$$

Hence if the cylinder be in contact with the plane, and a small velocity perpendicular to the plane be communicated to it, the cylinder will begin to move away from the plane with gradually increasing velocity. This velocity cannot however increase indefinitely, for that would require the energy to become infinite, which is impossible, since the energy remains constant and equal to its initial value. If R_0' denote the value of R' when the cylinder is in contact with the plane, v_0 the initial velocity, and v the velocity when the cylinder is at an infinite distance from the plane, the equation of energy gives

$$R_0'v_0^2 = R_\infty'v^2.$$

The value of $R_\infty' = M + M'$, since the motion is the same as if the plane boundary did not exist, whence the ratio of the terminal to the initial velocity is

$$\frac{v}{v_0} = \sqrt{\frac{R_0'}{M + M'}}.$$

Let us now suppose that the cylinder is projected parallel to the plane with initial velocity u_0. By (38) the initial acceleration $\dot{v}_0$ perpendicular to the plane is

$$\dot{v}_0 = \frac{u_0^2}{2R'}\frac{dR}{dy};$$

and since dR/dy is negative, the cylinder will be attracted towards the plane, and will ultimately strike it.

75. All the results of the last two sections are true in the case of a sphere, and can be proved in the same manner. Moreover the motion will be unaltered, if we remove the plane boundary, and suppose that on the other side, an infinite liquid exists in which

another equal cylinder or sphere is moving with velocities $u, -v$. The second cylinder or sphere is therefore the image of the first. Our results are therefore applicable to the case of two equal cylinders or spheres, which are moving with equal and opposite velocities along the line joining their centres; or to the case in which the cylinders or spheres are projected perpendicularly to the line joining their centres, with velocities which are equal and in the same direction.

These results have however a wider application, for according to the views of Faraday and Maxwell, the action which is observed to take place between electrified bodies, is not due to any direct action which electrified bodies exert upon one another, but to something which takes place in the dielectric medium surrounding these bodies; and although the preceding hydrodynamical results do not of course furnish any explanation of what takes place in dielectric media, they establish the fact that two bodies which are incapable of exerting any direct influence upon one another, are capable of producing an apparent attraction or repulsion upon one another, when they are in motion in a medium which may be treated as possessing the properties of an incompressible fluid.

EXAMPLES.

1. A light cylindrical shell whose cross section is an ellipse is filled with water, and placed at rest on a smooth horizontal plane in its position of unstable equilibrium. If it is slightly disturbed, prove that it will pass through its position of stable equilibrium with angular velocity ω, given by the equation

$$\omega^2 = \frac{8g(a^2 + b^2)}{(a+b)^2(a-b)}.$$

2. An elliptic cylindrical shell, the mass of which may be neglected, is filled with water, and placed on a horizontal plane very nearly in the position of unstable equilibrium with its axis horizontal, and is then let go. When it passes through the position of stable equilibrium, find the angular velocity of the cylinder, (i) when the horizontal plane is perfectly smooth, (ii) when it is

perfectly rough; and prove that in these two cases, the squares of the angular velocities are in the ratio

$$(a^2 - b^2)^2 + 4b^2(a^2 + b^2) : (a^2 - b^2)^2,$$

$2a$ and $2b$ being the axes of the cross section of the cylinder.

3. A pendulum with an elliptic cylindrical cavity filled with liquid, the generating lines of the cylinder being parallel to the axis of suspension, performs finite oscillations under the action of gravity. If l be the length of the equivalent pendulum, and l' the length when the liquid is solidified, prove that

$$l' - l = \frac{ma^2b^2}{h(M+m)(a^2+b^2)},$$

where M is the mass of the pendulum, m that of the liquid, h the distance of the centre of gravity of the whole mass from the axis of suspension, and a, b the semi-axes of the elliptic cavity.

4. Find the ratio of the kinetic energy of the infinite liquid surrounding an oblate spheroid, moving with given velocity in its equatorial plane, to the kinetic energy of the spheroid; and denoting this ratio by P, prove that if the spheroid swing as the bob of a pendulum under gravity, the distance between the axis of the suspension and the axis of the spheroid being c, the length of the simple equivalent pendulum is

$$\frac{(1+P)c + 2a^2/5c}{1 - \rho/\sigma},$$

where a is the equatorial radius, σ and ρ the densities of the spheroid and liquid respectively.

5. A pendulum has a cavity excavated within it, and this cavity is filled with liquid. Prove that if any part of the liquid be solidified, the time of oscillation will be increased.

6. A closed vessel filled with liquid of density ρ, is moved in any manner about a fixed point O. If at any time the liquid were removed, and a pressure proportional to the velocity potential were applied at every point of the surface, the resultant couple due to the pressure would be of magnitude G, and its direction in a line OQ. Show that the kinetic energy of the liquid was proportional to $\frac{1}{2}\rho\omega G \cos\theta$, where ω is the angular velocity of the surface, and θ the angle between its direction and OQ.

MOTION OF A SINGLE SOLID.

7. Liquid is contained in a simply-connected surface S; if ϖ is the impulsive pressure at any point of the liquid due to any arbitrary deformation of S, subject to the condition that the enclosed volume is not changed, and ϖ' the impulsive pressure for a different deformation, show that

$$\iint \varpi \frac{d\varpi'}{dn} dS = \iint \varpi' \frac{d\varpi}{dn} dS.$$

8. If a sphere be immersed in a liquid, prove that the kinetic energy of the liquid due to a given deformation of its surface, will be greater when the sphere is fixed than when it is free.

CHAPTER IV.

WAVES.

76. BEFORE discussing the dynamical theory of waves, we shall commence by explaining what a wave is.

Let us suppose that the equation of the free surface of a liquid at time t, is

$$y = a \sin (mx - nt) \quad \ldots\ldots\ldots\ldots\ldots\ldots (1),$$

where the axis of x is horizontal, the axis of y is measured vertically upwards, and a, m and n are constants.

The initial form of the free surface, i.e. its form when $t = 0$, is $y = a \sin mx$, which is the curve of sines. The maximum values of y occur when $mx = (2i + \frac{1}{2})\pi$, where i is zero or any positive or negative integer; and this maximum value is equal to a. The minimum values of y occur when $mx = (2i + \frac{3}{2})\pi$, where i is zero or any positive or negative integer; and this minimum value is equal to $-a$. As x increases from 0 to $\frac{1}{2}\pi/m$, y increases from 0 to a, and as x increases from $\frac{1}{2}\pi/m$ to π/m, y decreases from a to 0. As x increases from π/m to $\frac{3}{2}\pi/m$, y is negative; and when x has the latter value, y has attained its greatest negative value, which is equal to $-a$; as x increases from $\frac{3}{2}\pi/m$ to $2\pi/m$, y decreases from $-a$ to 0.

The values of y comprised between

$$x = 2i\pi/m \quad \text{and} \quad x = 2(i+1)\pi/m,$$

evidently go through exactly the same cycle of changes.

When the motion of a liquid is such, that its free surface is represented by an equation such as (1), the motion is called *wave motion*.

The quantity a, which is equal to the maximum value of y, is called the *amplitude*; and the distance $2\pi/m$, between two consecutive maxima values of y, is called the *wave length*.

In order to ascertain the form of the free surface at time t, let us transfer the origin to a point $\xi = nt/m$; then if x' be the abscissa at time t referred to the new origin, of the point whose abscissa referred to the old origin is x, we have $x = x' + \xi$ and

$$y = a \sin(mx' + m\xi - nt) = a \sin mx'.$$

The form of the free surface at time t, is therefore obtained by making the point which initially coincided with the origin, travel along the axis of x, with velocity n/m.

The velocity of this point is called the *velocity of propagation* of the wave.

If V be the velocity of propagation, and λ the wave length, we thus obtain the equations

$$m = 2\pi/\lambda, \quad V = n/m \quad\quad\quad\quad (2),$$

and therefore (1) may be written

$$y = a \sin \frac{2\pi}{\lambda}(x - Vt) \quad\quad\quad\quad (3).$$

If n, and therefore V, were negative, equations (1) and (3) would represent a wave travelling in the opposite direction.

The position of the free surface at time t, is exactly the same as at time $t + 2i\pi/n$, or $t + 2i\lambda/V$, since $n = 2\pi V/\lambda$; the quantity λ/V is called the *periodic time*, or shortly the *period*, and is equal to the time which the crest of one wave occupies in travelling from its position at time t, and the position occupied by the next crest at the same epoch. If τ denote the period, we evidently have

$$\lambda = V\tau \quad\quad\quad\quad (4),$$

and (1) may be written in the form

$$y = a \sin 2\pi \left(\frac{x}{\lambda} - \frac{t}{\tau}\right) \quad\quad\quad\quad (5),$$

which is a form sometimes convenient in Physical Optics.

From (4) we see that for waves travelling with the same velocity, the period increases with the wave length.

The reciprocal of the period, which is the number of vibrations executed per unit of time, is called the *frequency*. If therefore we

KINEMATICS OF WAVE MOTION. 87

have a medium which propagates waves of all lengths with the same velocity, equation (4) shows that the number of vibrations executed in a second, increases as the wave length diminishes. This remark is of importance in the Theory of Sound.

Let us now suppose that two waves are represented by the equations

$$y = a \sin \frac{2\pi}{\lambda} (x - Vt),$$

$$y' = a \sin \frac{2\pi}{\lambda} (x - Vt - e).$$

The amplitudes, wave lengths and velocities of propagation of the two waves are equal, but the second wave is in advance of the first; for if in the first equation we put $t + e/V$ for t, the two equations become identical. It therefore follows that the distance at which the second wave is in advance of the first, is equal to e. The quantity e is called the *phase* of the wave.

77. Waves which are represented by equations such as (1), are called *progressive waves*; their wave lengths are equal to $2\pi/m$, and their velocities of propagation to n/m. If such waves are travelling along the surface of water under the action of gravity, they may be conceived to have been produced by communicating to the free surface an initial displacement $y = a \sin mx$, together with an initial velocity $-an \cos mx$. We therefore see that the wave length depends solely on the initial displacement, but that the velocity of propagation depends upon the initial velocity as well as upon the initial displacement.

If we combine the two waves, which are obtained by writing $\pm n$ for n in (1) and adding the results, we shall obtain

$$y = a \sin (mx - nt) + a \sin (mx + nt)$$

$$= 2a \sin mx \cos nt \ldots\ldots\ldots\ldots\ldots\ldots\ldots\ldots (6).$$

Such a wave is called a *stationary wave*. It is produced by means of an initial displacement alone, and gives the form of the free surface at time t, when the latter is displaced into the form of the curve $y = 2a \sin mx$, and is then left to itself.

In equations (1) or (6), m, which is equal to $2\pi/\lambda$, is always supposed to be given, and the problem we have to solve, consists in finding the value of n, which determines the velocity of propagation.

78. We shall now proceed to consider the motion of irrotational liquid waves in two dimensions, under the action of gravity.

The solution of the problem involves the determination of a velocity potential ϕ, which satisfies the following three conditions:

(i) ϕ must satisfy Laplace's equation, and together with its first derivatives, must be finite and continuous at every point of the liquid.

(ii) ϕ must satisfy the given boundary conditions at the fixed boundaries of the liquid.

(iii) ϕ must be determined, so that the free surface of the liquid is a surface of constant pressure.

The difficulties of the subject are so great, that no rigorous solution of any problem has yet been obtained, except in the case of certain trochoidal waves, discovered by Gerstner, and which involve molecular rotation. If however the motion is sufficiently slow, terms involving the squares and products of the velocities may be neglected, and when this is the case, an approximate solution of a variety of problems can be obtained without difficulty.

We shall now find the condition to be satisfied at a free surface, when waves are propagated under the action of gravity.

Let the origin be taken in the undisturbed surface, and let the axis of x be measured in the direction of propagation of the waves, and let the axis of z be measured vertically upwards.

The pressure at any point of the liquid is determined by the equation
$$p/\rho + gz + \dot{\phi} + \tfrac{1}{2}q^2 = C \ldots\ldots\ldots\ldots\ldots (7).$$

The equation of the surfaces of constant pressure is $p = \text{const.}$, and since the free surface is included in this family of surfaces, *and must also satisfy the kinematical condition of a bounding surface*, it follows from § 12, (17) that
$$\frac{dp}{dt} + u\frac{dp}{dx} + v\frac{dp}{dy} + w\frac{dp}{dz} = 0 \ldots\ldots\ldots\ldots (8).$$

Substituting the value of p from (7) in (8), and neglecting squares and products of the velocity, we obtain
$$\frac{d^2\phi}{dt^2} + g\frac{d\phi}{dz} = 0 \ldots\ldots\ldots\ldots\ldots\ldots (9).$$

This is the condition to be satisfied at a free surface, where $z = 0$.

Waves in a Liquid of given Depth.

79. We shall now find the velocity of propagation of two-dimensional waves travelling in an ocean of depth h.

The equation of continuity is

$$\frac{d^2\phi}{dx^2} + \frac{d^2\phi}{dz^2} = 0 \quad \ldots\ldots\ldots\ldots\ldots\ldots (10).$$

The boundary condition at the bottom of the liquid is

$$\frac{d\phi}{dz} = 0, \text{ when } z = -h \quad \ldots\ldots\ldots\ldots (11).$$

To satisfy (10) assume

$$\phi = F(z) \cos(mx - nt) \quad \ldots\ldots\ldots\ldots (12).$$

Substituting in (10) we obtain

$$\frac{d^2 F}{dz^2} - m^2 F = 0,$$

the solution of which is

$$F = P \cosh mz + Q \sinh mz;$$

whence $\quad \phi = (P \cosh mz + Q \sinh mz) \cos(mx - nt)$.

Substituting in (11) and (9) we obtain

$$P \sinh mh = Q \cosh mh,$$
$$Pn^2 = Qmg;$$

whence eliminating P and Q, and taking account of (2), we obtain

$$V^2 = (g\lambda/2\pi) \tanh(2\pi h/\lambda) \quad \ldots\ldots\ldots\ldots (13),$$

which determines the velocity of propagation.

If the lengths of the waves are large in comparison with the depth of the liquid, h/λ is small, and the preceding result becomes

$$V^2 = gh \quad \ldots\ldots\ldots\ldots\ldots\ldots (14),$$

which determines the velocity of propagation of long waves in shallow water.

If the depth of the liquid is large in comparison with the wave length, h/λ is large, and $\tanh 2\pi h/\lambda = 1$ approximately, whence

$$V^2 = g\lambda/2\pi \quad \ldots\ldots\ldots\ldots\ldots (15),$$

which determines the velocity of propagation of deep sea waves.

The last result may be obtained directly, for the value of F may be written in the form
$$F = A\epsilon^{mz} + B\epsilon^{-mz},$$
and since ϕ, and therefore F, cannot be infinite when $z = -\infty$, $B = 0$, and (9) at once gives the required result.

80. Returning to the general case, we see that ϕ is of the form
$$\phi = A \cosh m(z+h) \cos(mx - nt).$$

If η be the elevation of the free surface above the undisturbed surface, we must have
$$\dot{\eta} = d\phi/dz \text{ when } z = 0 \dots\dots\dots\dots (16),$$
whence substituting the value of ϕ in (16), and suitably choosing the origin we obtain
$$\eta = -Amn^{-1} \sinh mh \sin(mx - nt).$$

Let (x, z) be the coordinates of an element of liquid when undisturbed, (ξ, ζ) its horizontal and vertical displacements, also let $x' = x + \xi$, $z' = z + \zeta$; then
$$\dot{\xi} = d\phi/dx' = -Am \cosh m(z'+h) \sin(mx' - nt)$$
$$\dot{\zeta} = d\phi/dz' = Am \sinh m(z'+h) \cos(mx' - nt).$$

Since the displacement is small we may put $x = x'$, $z = z'$ as a first approximation, and we obtain
$$\xi = -a \cosh m(z+h) \cos(mx - nt)$$
$$\zeta = -a \sinh m(z+h) \sin(mx - nt),$$
where $Am/n = a$; whence the elements of liquid describe the ellipse
$$\xi^2/\cosh^2 m(z+h) + \zeta^2/\sinh^2 m(z+h) = a^2.$$

When the depth of the liquid is very great we may put $h = \infty$, and the hyperbolic functions must be replaced by exponential ones; we shall thus obtain
$$\phi = A\epsilon^{mz} \cos(mx - nt)$$
$$\eta = -Amn^{-1} \sin(mx - nt),$$
and the elements of liquid will describe the circles
$$\xi^2 + \zeta^2 = (Am/n)^2 \epsilon^{mz}.$$

Waves at the Surface of Separation of Two Liquids.

81. Let us first suppose that two liquids of different densities (such as water and mercury) are resting upon one another, which are in repose except for the disturbance produced by the wave motion, and which are confined between two planes parallel to their surface of separation. Let ρ, ρ' be the densities of the lower and upper liquids respectively, h, h' their depths, and let the origin be taken in the surface of separation when in repose.

In the lower liquid let
$$\phi = A \cosh m(z+h) \cos(mx - nt),$$
and in the upper let
$$\phi' = A' \cosh m(z - h') \cos(mx - nt),$$
also let
$$\eta = a \sin(mx - nt)$$
be the equation of the surface of separation. At this surface, the condition that the two liquids should remain in contact requires that
$$d\eta/dt = d\phi/dz = d\phi'/dz, \text{ when } z = 0.$$
Whence $\quad -na = mA \sinh mh = -mA' \sinh mh'.$

If δp, $\delta p'$ be the increments of the pressure due to the wave motion just below and just above the surface of separation, then
$$\delta p + g\rho\eta + \rho d\phi/dt = 0,$$
and $\quad \delta p' + g\rho'\eta + \rho' d\phi'/dt = 0,$
and since $\delta p = \delta p'$, we obtain
$$g(\rho - \rho')\eta = -\rho d\phi/dt + \rho' d\phi'/dt$$
$$= n(-A\rho \cosh mh + A'\rho' \cosh mh') \sin(mx - nt)$$
$$= (\rho \coth mh + \rho' \coth mh') n^2\eta/m,$$
whence
$$U^2 = (n/m)^2 = \frac{g(\rho - \rho')}{m(\rho \coth mh + \rho' \coth mh')} \quad \ldots\ldots\ldots\ldots (17),$$
where $m = 2\pi/\lambda$.

82. When λ is small compared with h and h', then mh, mh' are large, and $\coth mh$ and $\coth mh'$ may be replaced by unity: we thus obtain
$$U^2 = g(\rho - \rho')/m(\rho + \rho').$$

If $\rho' > \rho$, U^2 is negative and therefore n is imaginary; hence if the upper liquid is denser than the lower, the motion cannot be represented by a periodic term in t, and is therefore unstable.

If the density of the upper liquid is small compared with that of the lower, we have approximately

$$U^2 = gm^{-1}(1 - 2\rho'/\rho).$$

If the liquid is water in contact with air, $\rho'/\rho = \cdot 00122$, hence if the air is treated as an incompressible fluid

$$U^2 = \cdot 99756 \times gm^{-1}.$$

83. Secondly, let us suppose that the upper liquid is moving with velocity V', and the lower with velocity V; then we may put

$$\phi = Vx + A \cosh m(z + h) \cos(mx - nt)$$
$$\phi' = V'x + A' \cosh m(z - h') \cos(mx - nt).$$

Let the equation of the surface of separation be

$$F = \eta - a \sin(mx - nt) = 0.$$

Then in both liquids F must be a bounding surface, and therefore by § 12 equation (17), when $z = 0$,

$$\frac{dF}{dt} + \frac{d\phi}{dx}\frac{dF}{dx} + \frac{d\phi}{dz}\frac{dF}{d\eta} = 0,$$

$$\frac{dF}{dt} + \frac{d\phi}{dx}\frac{dF}{dx} + \frac{d\phi'}{dz}\frac{dF}{d\eta} = 0.$$

Whence $\quad an - mVa + mA \sinh mh = 0$

$\qquad\qquad an - mV'a - mA' \sinh mh' = 0.$

Hence if $U = n/m$ be the velocity of propagation,

$$A \sinh mh = a(V - U)$$
$$A' \sinh mh' = -a(V' - U).$$

If δp, $\delta p'$ be the increments of pressure at the surface of separation due to the wave motion

$$\delta p/\rho + g\eta + d\phi/dt + \tfrac{1}{2}\{V - Am \cosh mh \cos(mx - nt)\}^2 = \tfrac{1}{2}V^2,$$
$$\delta p'/\rho' + g\eta + d\phi'/dt + \tfrac{1}{2}\{V' - A'm \cosh mh' \cos(mx - nt)\}^2 = \tfrac{1}{2}V'^2.$$

Therefore since $\delta p = \delta p'$,

$$ag(\rho - \rho') = Amp(V - U)\cosh mh - A'm\rho'(V' - U)\cosh mh'$$

or $\quad g(\rho - \rho') = m\rho(V - U)^2 \coth mh + m\rho'(V' - U)^2 \coth mh' \ldots(18)$

which determines U.

Stability and Instability.

84. We shall now consider a question which has excited a good deal of attention of late years, viz. the stability or instability of fluid motion.

If a disturbance be communicated to the two liquids which are considered in §§ 81—83, the surface of separation may be conceived to be initially of the form $\eta = a \sin mx$ or $a \cos mx$, where m is a given real quantity, whose value depends upon the nature of the disturbance. An equation of this kind does not of course represent the most general possible kind of disturbance, but in as much as by a general theorem due to Fourier, any arbitrary function can be expressed in the form of a series of sines or cosines, or by a definite integral involving such quantities, an equation of this form is sufficient for our purpose.

We have pointed out that the object of the wave motion problem is to determine n; if therefore n should be found to be a real quantity, the subsequent motion will be periodic, and therefore stable; but if n should turn out to be an imaginary or complex quantity, the final solution will involve real exponential quantities, and therefore the motion will increase indefinitely with the time and will be unstable.

To understand this more clearly let us employ complex quantities, and assume that the initial form of the free surface is the real part of

$$\eta = (A - \iota B) e^{\iota mx},$$

where A, B and m are real; and let n be of the form $\alpha + \iota \beta$. Since the form of the free surface at any subsequent time is

$$\eta = (A - \iota B) e^{\iota(mx - nt)},$$

this becomes $\eta = (A - \iota B) e^{\iota(mx - \alpha t) + \beta t},$

the real part of which is

$$\eta = e^{\beta t} \{A \cos (mx - \alpha t) + B \sin (mx - \alpha t)\} \ldots \ldots (19).$$

If therefore β is positive, the amplitude will increase indefinitely with the time, and the motion will be unstable. In such a case the two liquids will, after a short time, become mixed together, and will usually remain permanently mixed, if they are capable of mixing; but if they are incapable of remaining permanently mixed, the lighter liquid will gradually work its way upwards, and a stable condition will ultimately be arrived at.

85. If one liquid is resting upon another, equilibrium is possible when the heavier liquid is at the top, but in this case the equilibrium is unstable; for since $\rho' > \rho$, it follows from (17) that n^2 is negative and therefore n is of the form $\pm \iota\beta$. Hence in the beginning of the disturbed motion, the free surface is of the form
$$\eta = A\epsilon^{\pm \beta t} \cos(mx - e).$$

If the upper liquid is moving with velocity V', and the lower with velocity V, the values of U or n/m are determined by the quadratic (18); and the condition of stability requires that the two roots of this quadratic should be real.

Putting k, k' for $m \coth mh$ and $m \coth mh'$, (18) becomes
$$k\rho (V - U)^2 + k'\rho' (V' - U)^2 = g(\rho - \rho').$$

The condition that the roots of this quadratic in U should be real, is
$$g(k\rho + k'\rho')(\rho - \rho') - kk'\rho\rho'(V - V')^2 > 0.$$

It therefore follows that if $\rho > \rho'$, that is if the lower liquid is denser than the upper liquid, the motion *may be stable*. But if $\rho' > \rho$; or if no forces are in action, so that $g = 0$, the motion will be unstable.

86. If no forces are in action, and both liquids are of unlimited extent so that $h = h' = \infty$, the equation for determining U becomes
$$\rho (V - U)^2 + \rho'(V' - U)^2 = 0,$$
the roots of which are
$$U = \frac{\rho V + \rho' V' \pm \iota \sqrt{\rho \rho'}\,(V - V')}{\rho + \rho'} \quad \ldots\ldots\ldots\ldots(20).$$

Hence U, and therefore n, is a complex quantity, and we may therefore put
$$U = \alpha \pm \iota\beta = n/m,$$
where α and β are determined from (20). If therefore the initial form of the free surface is
$$\eta = a\epsilon^{\iota m x},$$
its form at any subsequent time may be written
$$\eta = \epsilon^{\iota m(x - \alpha t)} \{a' \epsilon^{m\beta t} + b' \epsilon^{-m\beta t}\} \ldots\ldots\ldots\ldots(21),$$
where $a' + b' = a$.

If there is no initial displacement, $\dot\eta = 0$ when $t = 0$, in which case $a' = b' = \frac{1}{2}a$. To express this result in real quantities, let $a = A - \iota B$, and (21) becomes
$$\eta = \{A \cos m(x - \alpha t) + B \sin m(x - \alpha t)\} \cosh m\beta t,$$
corresponding to an initial displacement
$$\eta = A \cos mx + B \sin mx.$$

87. When the initial velocity is zero there are three cases worthy of notice.

(i) Let $\rho = \rho'$, $V = -V'$, so that the densities of the two liquids are equal, and their undisturbed velocities are equal and opposite; then from (20), $\alpha = 0$, $\beta = V$, whence
$$\eta = (A \cos mx + B \sin mx) \cosh mVt.$$

(ii) Let $\rho = \rho'$, $V' = 0$, then $\alpha = \frac{1}{2}V$, $\beta = \frac{1}{2}V$, whence
$$\eta = \{A \cos m(x - \tfrac{1}{2}Vt) + B \sin m(x - \tfrac{1}{2}Vt)\} \cosh \tfrac{1}{2}mVt,$$
hence the waves travel in the direction of the stream and with half its velocity.

(iii) Let $\rho = \rho'$, $V = V'$. In this case the roots are equal, but the general solution may be obtained from (20) by putting $V' = V(1 + \gamma)$, where γ ultimately vanishes. We thus obtain
$$\alpha = V + \tfrac{1}{2}V\gamma, \quad \beta = -\tfrac{1}{2}V\gamma,$$
and therefore since γ is small, (21) may be written
$$\eta = \epsilon^{\iota m(x - Vt)}\{a + \tfrac{1}{2}mV\gamma t [a(1 - \iota) - 2a']\}.$$
Putting $c = \tfrac{1}{2}mV\gamma\{a(1 - \iota) - 2a'\}$, this becomes
$$\eta = (a + ct)\epsilon^{\iota m(x - Vt)}.$$
Let $a = A - \iota B$, $c = C - \iota D$, then the real part is
$$\eta = (A + Ct) \cos m(x - Vt) + (B + Dt) \sin m(x - Vt),$$
corresponding to an initial displacement $\eta = A \cos mx + B \sin mx$. If the initial velocity $\dot\eta$ is zero, $C = mBV$, $D = -mAV$, and
$$\eta = (A + mBVt) \cos m(x - Vt) + (B - mAVt) \sin m(x - Vt).$$

The peculiarity of this solution is, that previously to displacement there is no real surface of separation at all. Hence if we have a thin surface such as a flag, whose inertia may be neglected, dividing the air, it appears from the last equation that (neglecting changes in the density of the air), the motion of the flag will be unstable and that it will flap.

Long Waves in Shallow Water.

88. In the theory of long waves it is assumed, that the lengths of the waves are so great in proportion to the depth of the water, that the vertical component of the velocity can be neglected, and that the horizontal component is uniform across each section of the canal. In § 79 we saw that if the depth is small compared with the wave length, then $U^2 = gh$, provided the square of the velocity is neglected. We shall now examine this result in connection with the above-mentioned assumption.

Let the motion be made steady by impressing on the whole liquid a velocity equal and opposite to the velocity of propagation of the waves. Let η be the elevation of the liquid above the undisturbed surface; U, u the velocities corresponding to h and $h + \eta$ respectively. The equation of continuity gives

$$u = hU/(h + \eta),$$

whence $\qquad U^2 - u^2 = U^2(2h\eta + \eta^2)/(h + \eta)^2.$

If δp be the excess of pressure due to the wave motion

$$\delta p = \left\{ \frac{U^2(2h + \eta)}{2(h + \eta)^2} - g \right\} \rho\eta.$$

When η/h is very small, the quantity in brackets is $U^2/h - g$; whence if $U^2 = gh$, the change of pressure at a height $h + \eta$ vanishes to a first approximation, and therefore a free surface is possible.

If the condition $U^2 = gh$ is satisfied, the change of pressure to a second approximation is

$$\delta p = -3g\rho\eta^2/2h,$$

which shows that the pressure is defective at all parts of the wave at which η differs from zero. *Unless therefore η^2 can be neglected, it is impossible to satisfy the condition of a free surface for a stationary long wave;—in other words, it is impossible for a long wave of finite height to be propagated in still water without change of type.* If however η be everywhere positive, a better result can be obtained with a somewhat increased value of U; and if η be everywhere negative, with a diminished value. We therefore infer that positive waves travel with a somewhat higher, and negative waves with a somewhat lower velocity than that due to half the undisturbed depth[1].

[1] Lord Rayleigh, "On Waves," *Phil. Mag.* April, 1876.

89. The theory of long waves in a canal may be investigated analytically as follows[1].

Let the origin be in the bottom of the liquid, h the undisturbed depth, η the elevation; and let x be the abscissa of an element of liquid when undisturbed, ξ the horizontal displacement. The quantity of liquid originally between the planes x and $x + dx$ is hdx; at the end of an interval t, the breadth of this stratum is $dx(1 + d\xi/dx)$, and its height is $h + \eta$, whence the equation of continuity is

$$(1 + d\xi/dx)(h + \eta) = h \quad\ldots\ldots\ldots\ldots(22).$$

Let us now investigate the motion of a column of liquid contained between the planes whose original distance was dx; and let us suppose that in addition to gravity, small horizontal and vertical disturbing forces X and Y act. Since the vertical acceleration is neglected, the pressure will be equal to the hydrostatic pressure due to a column of liquid of height $h + \eta$, whence

$$p = g\rho(h + \eta - y) + \rho \int_y^{h+\eta} Y dy \ldots\ldots\ldots(23).$$

The equation of motion of the stratum is

$$\rho h \frac{d^2\xi}{dt^2} = -\frac{dp}{dx}(h + \eta) + X\rho h \ldots\ldots\ldots\ldots(24).$$

Now from (23),

$$\frac{dp}{dx} = g\rho \frac{d\eta}{dx} + \rho Y \frac{d\eta}{dx} + \rho \int_y^{h+\eta} \frac{dY}{dx} dy \ldots\ldots\ldots(25);$$

also in most problems to which the theory applies, the last two terms on the right-hand side of (25) are very much smaller than the first, and may therefore be neglected, whence (24) becomes

$$h \frac{d^2\xi}{dt^2} = -g(h + \eta)\frac{d\eta}{dx} + Xh.$$

Substituting the value of η from (22) we obtain

$$\frac{d^2\xi}{dt^2} = gh \frac{d^2\xi}{dx^2}\left(1 + \frac{d\xi}{dx}\right)^{-3} + X \quad\ldots\ldots\ldots\ldots(26).$$

For a first approximation, we may neglect squares and products of small quantities, and (22) and (26) respectively become

$$\eta/h = -d\xi/dx \ldots\ldots\ldots\ldots\ldots(27),$$

$$\frac{d^2\xi}{dt^2} = gh \frac{d^2\xi}{dx^2} + X \ldots\ldots\ldots\ldots\ldots(28).$$

[1] Airy, "Tides and Waves," *Encyc. Met.*

98   WAVES.

In order to solve (28) when $X = 0$, assume $\xi = \epsilon^{\iota(mx-nt)}$, and we obtain $n/m = (gh)^{\frac{1}{2}}$, which shows that the velocity of propagation is equal to $(gh)^{\frac{1}{2}}$.

Stationary Waves in Flowing Water[1].

90. Let us suppose that water is flowing uniformly along a straight canal with vertical sides, and that between two points A and B there are small inequalities, and that beyond these points the bottom is perfectly level. Let a be the depth, u the velocity, p the *mean* pressure beyond A; b the depth, v the velocity, and q the *mean* pressure beyond B: also let f be the difference of levels of the bottom at A and B.

The total energy of the liquid per unit of the canal's length and breadth, at points beyond B, is

$$\tfrac{1}{2}v^2 b + g\int_0^b y\, dy + w = \tfrac{1}{2}(v^2 + gb)b + w,$$

where w is the wave energy, and the density of the liquid is taken as unity. At very great distances beyond B the wave motion will have subsided and w will be zero.

The equation of continuity is

$$au = bv = M \quad\ldots\ldots\ldots\ldots\ldots\ldots(29).$$

The dynamical equation is found from the consideration, that the difference between the work done by the pressure p upon the volume of water entering at A, and the work done by the pressure q at B upon an equal volume of water passing away at B, is equal to the difference between the energy which passes away at B, and the energy which enters at A. Whence

$$pau - qbv = (\tfrac{1}{2}v^2 b + \tfrac{1}{2}gb^2 + w)v - (\tfrac{1}{2}u^2 a + g\int_f^{a+f} y\, dy)u,$$

which by (29) becomes,

$$p - q = \tfrac{1}{2}v^2 + \tfrac{1}{2}gb + w/b - \tfrac{1}{2}u^2 - g(f + \tfrac{1}{2}a)\ldots\ldots\ldots(30).$$

Now p and q are the *mean* pressures, and therefore since the pressure at the free surface is zero,

$$p = \tfrac{1}{2}ga, \quad q = \tfrac{1}{2}gb + w'/b,$$

[1] Sir W. Thomson, *Phil. Mag.* (5) vol. xxii. 353.

where w' denotes a quantity depending on the wave disturbance; whence (30) becomes

$$\tfrac{1}{2}M^2(a^2-b^2)/a^2b^2 - g(a-b+f) + (w-w')/b = 0\ldots\ldots(31).$$

If we put
$$D^3 = 2a^2b^2/(a+b), \quad M = VD;$$

D will denote a mean depth intermediate between a and b, and approximately equal to their arithmetic mean when their difference is small in comparison with either; and V will similarly denote a corresponding mean velocity of flow. We thus obtain from (31)

$$b - a = \frac{f - (w-w')/gb}{1 - V^2/gD}.$$

If $b - a$ were exactly equal to f, and there were no disturbance of the water beyond B, the mean level of the water would be the same at great distances beyond A and B; but if this is not the case, there will be a rise or fall of level, determined by the formula

$$y = b - a - f = \frac{V^2 f/gD + (w-w')/gb}{1 - V^2/gD}.$$

Let us now suppose that between A and B there are various small inequalities; each of these inequalities will produce small waves whose nature is determined by the form of the functions w, w'; hence w and w' will both be small quantities and the sign of y will be independent of that of $w - w'$. Now f is positive or negative according as the bottom at A is higher or lower than the bottom at B. *Hence if $V^2 < gD$ the upper surface of the water rises when the bottom falls, and falls when the bottom rises; and the converse is the case when $V^2 > gD$.*

Theory of Group Velocity.

91. When a group of waves advances into still water, it is observed that the velocity of the group is less than that of the individual waves of which it is composed. This phenomenon was first explained by Sir G. Stokes[1], who regarded the group as formed by the superposition of two infinite trains of waves of equal amplitudes and nearly equal wave lengths, advancing in the same direction.

[1] *Smith's Prize Examination*, 1876; and Lord Rayleigh, "On Progressive Waves"; *Proc. Lond. Soc.* vol. IX.

Let the two trains of waves be represented by $\cos k\,(Vt - x)$ and $\cos k'\,(V't - x)$; their resultant is equal to

$$\cos k\,(Vt - x) + \cos k'\,(V't - x) = 2\cos\tfrac{1}{2}\{(k'V' - kV)\,t - (k' - k)\,x\}$$
$$\times \cos\tfrac{1}{2}\{(k'V' + kV)\,t - (k' + k)\,x\}.$$

If $k' - k$, $V' - V$ be small, this represents a train of waves whose amplitude varies slowly from one point to another between the limits 0 and 2, forming a series of groups separated from one another by regions comparatively free from disturbance. The position at time t of the middle of the group, which was initially at the origin, is given by

$$(k'V' - kV)\,t - (k' - k)\,x = 0,$$

which shows that the velocity of propagation U of the group is

$$U = (k'V' - kV)/(k' - k).$$

In the limit when the number of waves in each group is indefinitely great we have $k' = k + \delta k$, $V' = V + \delta V$, whence

$$U = \frac{d\,(kV)}{dk}.$$

Capillary Waves.

92. Most liquids which are incapable of remaining permanently mixed, exhibit a certain phenomenon called capillarity[1], when in contact with one another. This phenomenon can be explained, by supposing that the surface of separation is capable of sustaining a tension, which is equal in all directions, and is independent of the form of the surface of separation.

The surface tension depends upon the nature of both the liquids which are in contact with one another. Thus at a temperature of 20° C., the surface tension of water in contact with air is 81 dynes per centimetre; whilst the surface tension of water in contact with mercury is 418 dynes per centimetre.

The surface tension diminishes as the temperature increases; also a surface tension cannot exist at the common surface of two

---

[1] The reader who desires to study the theory of Capillarity is recommended to consult Chapter xx. of Maxwell's *Heat*; and also the article on Capillarity in the *Encyclopædia Britannica* by the same author.

A table of the superficial tensions of various liquids will be found in Everett's *Units and Physical Constants*, p. 49.

liquids, such as water and alcohol, which are capable of becoming permanently mixed.

93. We shall now consider the effect of surface tension upon the propagation of waves.

Let T be the surface tension, and let p and $p + \delta p$ be the pressures just outside and just inside the free surface of a liquid; then

$$\delta p/\rho + g\eta + \dot{\phi} = 0 \quad\quad\quad\quad\quad (32).$$

But if we resolve the forces which act upon a small element δs of the free surface vertically, and neglect the vertical acceleration, and put $\delta\chi$ for the angle which δs subtends at the centre of curvature, we obtain

$$\delta p \delta s = T \delta \chi,$$

whence
$$\delta p = T \frac{d\chi}{ds}.$$

Now
$$\frac{d\eta}{dx} = \cot \chi,$$

therefore
$$\frac{d^2\eta}{dx^2} = -\operatorname{cosec}^2 \chi \frac{d\chi}{ds}.$$

Since χ is nearly equal to $\tfrac{1}{2}\pi$, we may put $\operatorname{cosec} \chi = 1$, and $ds = dx$, whence

$$\delta p = - T \frac{d^2\eta}{dx^2}.$$

Substituting in (32), differentiating the result with respect to t, and remembering that $\dot{\eta} = d\phi/dz$, and that $d^2\phi/dx^2 = - d^2\phi/dz^2$, we obtain

$$\frac{T}{\rho} \frac{d^3\phi}{dz^3} + g \frac{d\phi}{dz} + \frac{d^2\phi}{dt^2} = 0 \quad\quad\quad\quad\quad (33).$$

This is the condition to be satisfied at the free surface.

94. We shall now apply the preceding result to determine the capillary waves propagated in an ocean of depth h.

Let $\quad \phi = A \cosh m (z + h) \cos (mx - nt).$

Substituting in (33) we obtain

$$(Tm^3/\rho + mg) \sinh mh = n^2 \cosh mh,$$

whence

$$U^2 = n^2/m^2 = (g\lambda/2\pi + 2\pi T/\rho\lambda) \tanh 2\pi h/\lambda \quad\quad\quad (34).$$

Equation (34) determines the velocity of propagation corresponding to a given wave length.

Let us now suppose that the depth of the liquid is so great, that $\tanh 2\pi h/\lambda$ may be replaced by unity. Equation (34) becomes

$$g\rho\lambda^2 - 2\pi\rho U^2\lambda + 4\pi^2 T = 0\ldots\ldots\ldots\ldots(35),$$

whence $\quad \lambda = \pi U^2/g \pm \pi g^{-1}\sqrt{(U^4 - 4Tg/\rho)}.$

In order that wave motion may be possible both values of λ must be real, which requires that

$$U = \text{or} > (4Tg/\rho)^{\frac{1}{4}}.$$

Hence the minimum value of U is $(4Tg/\rho)^{\frac{1}{4}}$, and the corresponding value of λ is $2\pi\sqrt{(T/g\rho)}$.

Sir W. Thomson[1] defines a *ripple* to be a wave whose length is less than the preceding critical value of λ.

95. In § 80 we have considered the propagation of waves at the surface of separation of two liquids, which are moving with different velocities. We shall now consider the production of ripples by wind blowing over the surface of still water.

Let V be the velocity of the wind, which is supposed to be parallel to the undisturbed surface of the water, σ the density of air referred to water.

Since the changes of density of the air are very small in the neighbourhood of the water, the air may approximately be regarded as an incompressible fluid, whence if the accented letters refer to the water, the kinematical conditions at the boundary give

$$\phi = Vx + a(U - V)\epsilon^{-mz}\cos(mx - nt),$$

$$\phi' = -aU\epsilon^{mz}\cos(mx - nt),$$

where U is the velocity of propagation of the waves in the water, and $\eta = a\sin(mx - nt)$ is the equation of its free surface.

Since the vertical acceleration is neglected the dynamical condition at the free surface is

$$T\delta\chi + (\delta p - \delta p')\delta s = 0,$$

or $\quad\quad \delta p - \delta p' = T\dfrac{d^2\eta}{dx^2}\ldots\ldots\ldots\ldots\ldots(36).$

Now

$$\delta p + g\sigma\eta + \dot\phi + \tfrac{1}{2}\left\{V - am(U - V)\sin(mx - nt)\right\}^2 - \tfrac{1}{2}V^2 = 0,$$

or $\quad \delta p + a\sigma\left\{g + n(U - V) - m(U - V)\right\}\sin(mx - nt) = 0.$

[1] *Phil. Mag.* (4), vol. XLII.

Similarly
$$\delta p' + (g - Un) a \sin(mx - nt) = 0,$$
whence (36) becomes
$$g(\sigma - 1) + \sigma m (U - V)^2 + mU^2 - Tm^2 = 0 \quad \ldots\ldots(37).$$

Let W be the velocity of propagation of waves in water when there is no wind, then
$$W = \sqrt{\frac{g(1-\sigma) + Tm^2}{m(1+\sigma)}},$$
or
$$Tm^2 - m(1+\sigma)W^2 + g(1-\sigma) = 0\ldots\ldots\ldots\ldots(38).$$

The condition that the roots of this quadratic in m should be real is that
$$W^2 = \text{or} > \frac{2}{1+\sigma}\sqrt{Tg(1-\sigma)} \ldots\ldots\ldots\ldots\ldots(39),$$
which determines the minimum value of W. This value of W is less than $(4Tg)^{\frac{1}{4}}$, which shows that when water is in contact with air it is possible for ripples to travel over its surface.

Substituting the value of T from (38) in (37) we obtain
$$U^2 - 2\sigma VU + \sigma V^2 - (1+\sigma)W^2 = 0,$$
whence
$$U = \frac{\sigma V}{1+\sigma} \pm \sqrt{\left\{W^2 - \frac{\sigma V^2}{(1+\sigma)^2}\right\}} \ldots\ldots\ldots\ldots(40).$$

We shall now discuss this equation.

Case (i). $V < W\sqrt{(1+\sigma)/\sigma}$.

In this case both values of U are real, and one of them is positive and the other negative; hence waves can travel either with or against the wind. Moreover since the positive value is numerically greater than the negative value, waves travel faster with the wind than against the wind; also the velocity of waves travelling against the wind is always less than W.

Case (ii). $V > W\sqrt{(1+\sigma)/\sigma}$.

In this case both values of U if real, are positive; hence waves cannot travel against the wind.

Case (iii). When $V < 2W$, the velocity of waves travelling with the wind is $> W$; when $V > 2W$ this velocity is $< W$; and when $V = 2W$, the velocity of waves travelling with the wind is undisturbed.

Case (iv). If $V > W(1+\sigma)\sigma^{-\frac{1}{2}}$, both values of U are imaginary, which shows that the motion is unstable.

EXAMPLES.

1. A liquid of infinite depth is bounded by a fixed plane perpendicular to the direction of propagation of the waves. Prove that each element of liquid will vibrate in a straight line, and draw a figure representing the free surface and the direction of motion of the elements, when the crest of the wave reaches the fixed plane.

2. Prove that the velocity of propagation of long waves in a semicircular canal of radius a and whose banks are vertical, is
$$\tfrac{1}{2}(\pi g a)^{\frac{1}{2}}.$$

3. If two series of waves of equal amplitude and nearly equal wave-length travel in the same direction, so as to form alternate lulls and roughness, prove that in deep water these are propagated with half the velocity of the waves; and that as the ratio of the depth to the wave length decreases from ∞ to 0, the ratio of the two velocities of propagation increases from $\tfrac{1}{2}$ to 1.

4. If a small system of rectilinear waves move parallel to and over another large rectilinear system, prove that the path of a particle of water is an epicycloid or hypocycloid, according as the two systems are moving in the same or opposite directions.

5. A fine tube made of a thin slightly elastic substance is filled with liquid; prove that the velocity of propagation of a disturbance in the liquid is $(\lambda \theta / a \rho)^{\frac{1}{2}}$, where a is the internal diameter of the tube, θ its thickness, λ the coefficient of elasticity of the material of which it is made, and ρ the density of the liquid.

6. A horizontal rectangular box is completely filled with three liquids which do not mix, whose densities reckoned downwards are σ_1, σ_2, σ_3, and whose depths when in equilibrium are l_1, l_2, l_3 respectively. Show that if long waves are propagated at their common surfaces, the velocity of propagation V must satisfy the equation

$$\{(\sigma_1/l_1 + \sigma_2/l_2)V^2 - g(\sigma_2 - \sigma_1)\}\{(\sigma_2/l_2 + \sigma_3/l_3)V^2 - g(\sigma_3 - \sigma_2)\} = \sigma_2^2 V^4/l_2^2.$$

7. Prove that liquid of density ρ flowing with mean velocity U through an elastic tube of radius a, will throw the surface into slight stationary corrugations, of which the number per unit of length is
$$(2\rho a U^2 - \lambda)^{\frac{1}{2}}/(2\pi a T)^{\frac{1}{2}},$$
where λ is the modulus of elasticity of the substance of the tube, and T its total tension.

8. Prove that the velocity potential
$$\phi = A\,(\lambda + 2\pi^2 y^2/\lambda)\sin 2\pi\,(vt - x)/\lambda$$
satisfies the equation of continuity in a mass of water, provided the ratio y/λ is so small for all possible values of y, that its square may be neglected. Hence prove that if the water in a canal of uniform breadth and uniform depth k, be acted upon in addition to gravity by the horizontal force $Ha^{-1}\sin 2\,(mt - x/a)$, where H and m are small and a is large, the equation of the free surface may be of the form
$$y = k + \frac{Hk}{2\,(gk - m^2 a^2)}\cos 2\,(mt - x/a).$$

9. Two liquids of density ρ, ρ' completely fill a shallow pipe; prove that the velocity of propagation of long waves is
$$U^2 = \frac{g\,(\rho - \rho')\,AA'}{b\,(A'\rho + A\rho')},$$
where A, A' are the areas of the vertical sections of the two liquids when undisturbed, and b is the breadth of the surface of separation.

10. If the upper liquid were moving with mean velocity U', and the lower with mean velocity U, and there is a surface tension T, prove that the wave-length is determined by the equation
$$4T\pi^2/\lambda^2 = b\,(\rho U^2/A + \rho' U'^2/A') - g\,(\rho - \rho').$$

11. If the bottom of a horizontal canal of depth h be constrained to execute a simple harmonic motion, such that the vertical displacement at a distance x from a given line across the canal and perpendicular to its length, be given by $k\cos m\,(x - vt)$, k being small; show that when the motion is steady, the form of the free surface is given by
$$y = h + \frac{kv^2}{v^2 - gh}\cos m\,(x - vt).$$

12. A shallow trough is filled with oil and water, the depth of the water being k and its density σ, and that of the oil being h and its density ρ. Prove that the velocity of propagation v of long waves is

$$v^2/g = \tfrac{1}{2}(h+k) + \tfrac{1}{2}\{(h-k)^2 + 4hk\rho/\sigma\}^{\tfrac{1}{2}}.$$

(Note that there may be slipping between the oil and water.)

13. If water is flowing with velocity proportional to the distance from the bottom, V being the velocity of the stream at its surface, prove that the velocity of propagation U of waves in the direction of the stream is given by

$$(U-V)^2 - V(U-V)\,W^2/gh - W = 0,$$

where W is the velocity of propagation of waves in still water.

14. Two liquids of densities ρ, ρ', each of which half fills a pipe of which the cross section is a square with a vertical diagonal of length $2h$, are slightly disturbed. Neglecting the disturbing effect of the boundary in the neighbourhood of the surface of separation, prove that the velocity of propagation of progressive waves along the pipe is given by the equation

$$U^2 = \frac{g(\rho-\rho')}{2m(\rho+\rho')}(\tanh \text{ or } \coth)\,mh.$$

CHAPTER V.

RECTILINEAR VORTEX MOTION.

96. THE present Chapter will be devoted to the consideration of certain problems of two-dimensional motion, which involve molecular rotation.

A vortex line may be defined to be a line whose direction coincides with the direction of the instantaneous axis of molecular rotation. Hence the differential equations of a vortex line are

$$\frac{dx}{\xi} = \frac{dy}{\eta} = \frac{dz}{\zeta}.$$

When the motion is in two dimensions,

$$w = 0, \quad du/dz = 0, \quad dv/dz = 0, \quad \xi = 0, \quad \eta = 0,$$

and therefore the vortex lines are all parallel to the axis of z.

In the case of a liquid, it follows from § 18, equation (26), that the rotation ζ may be any function of x, y and t, which satisfies the equation,

$$\frac{d\zeta}{dt} + u\frac{d\zeta}{dx} + v\frac{d\zeta}{dy} = 0,$$

and this is satisfied by $\zeta = $ const., we shall therefore suppose that ζ is always constant at every point of the liquid where molecular rotation exists.

97. We have already shown that when the motion is in two dimensions, a current function always exists, such that

$$u = d\psi/dy, \quad v = -d\psi/dx;$$

also u, v and ζ are connected together by the equation

$$\frac{dv}{dx} - \frac{du}{dy} = 2\zeta \quad \dots\dots\dots\dots(1),$$

whence substituting for u, v in terms of ψ, we obtain

$$\frac{d^2\psi}{dx^2} + \frac{d^2\psi}{dy^2} + 2\zeta = 0 \quad \dots\dots\dots\dots(2).$$

This equation must be satisfied at every point of the liquid where vortex motion exists. At every point of the irrotationally moving liquid which surrounds the vortices, $\zeta = 0$, and therefore

$$\frac{d^2\psi}{dx^2} + \frac{d^2\psi}{dy^2} = 0 \quad \dots\dots\dots\dots(3).$$

Equations (2) and (3) show, that ψ is the potential of indefinitely long cylinders composed of attracting matter of density $\zeta/2\pi$, which occupy the same positions as the vortices.

Let us now suppose that a single rectilinear vortex, whose cross section is a circle of radius a, exists in an infinite liquid. In order that the cross section may remain circular, it is necessary that ψ should be a function of r alone.

Denoting the values of quantities inside the vortex by accented letters, equations (2) and (3) become

$$\frac{d^2\psi'}{dr^2} + \frac{1}{r}\frac{d\psi'}{dr} + 2\zeta = 0 \quad \dots\dots\dots\dots(4),$$

which gives the values of ψ inside the vortex, and

$$\frac{d^2\psi}{dr^2} + \frac{1}{r}\frac{d\psi}{dr} = 0 \quad \dots\dots\dots\dots(5),$$

which gives the value outside.

The complete integrals of (4) and (5) are

$$\psi' = A \log r + B - \tfrac{1}{2}\zeta r^2$$

and $$\psi = C \log r + D.$$

Now ψ' must not be infinite when $r = 0$, and therefore $A = 0$; also at the boundary of the vortex, where $r = a$.

$$\psi' = \psi, \qquad d\psi'/dr = d\psi/dr;$$

whence $$B - \tfrac{1}{2}\zeta a^2 = C \log a + D$$

$$-\zeta a^2 = C,$$

and therefore $$C = -\zeta a^2 = -\zeta \sigma/\pi = -m/\pi,$$

SINGLE CIRCULAR VORTEX.   109

where σ is the area of the cross section, and m is the *strength* of the vortex. The constant D contributes nothing to the velocity, and may therefore be omitted, whence

$$\psi' = \tfrac{1}{2}\zeta(a^2 - r^2) - (m/\pi)\log a \quad\ldots\ldots\ldots\ldots\ldots(6),$$

$$\psi = -(m/\pi)\log r \quad\ldots\ldots\ldots\ldots\ldots\ldots\ldots\ldots(7).$$

Now $-d\psi/dr$ is the velocity perpendicular to r, whence inside the vortex

$$-d\psi'/dr = \zeta r \quad\ldots\ldots\ldots\ldots\ldots\ldots\ldots(8),$$

which vanishes when $r = 0$, and outside

$$-d\psi/dr = m/\pi r \quad\ldots\ldots\ldots\ldots\ldots\ldots(9).$$

Hence a single vortex whose cross section is circular, if existing in an infinite liquid will remain at rest, and will rotate as a rigid body. It will also produce at every point of the irrotationally moving liquid with which it is surrounded, a velocity which is perpendicular to the line joining that point with the centre of its cross section, and which is inversely proportional to the distance of that point from the centre.

98. Outside the vortex, where the motion is irrotational, a velocity potential of course exists. To find its value we have

$$\frac{d\phi}{dx} = \frac{d\psi}{dy} = -\frac{my}{\pi r^2}, \quad \frac{d\phi}{dy} = -\frac{d\psi}{dx} = \frac{mx}{\pi r^2},$$

whence $\quad \phi = -\dfrac{m}{\pi}\displaystyle\int \dfrac{ydx - xdy}{x^2 + y^2} = (m/\pi)\tan^{-1}y/x \ \ldots\ldots(10).$

It therefore follows that ϕ is a many valued function, whose cyclic constant is $2m$. The circulation, i.e. the line integral $\int(udx + vdy)$, is zero when taken round any closed curve which does not surround the vortex, and is equal to $2m$, when the curve surrounds the vortex; whence if κ be the circulation, $m = \tfrac{1}{2}\kappa$, and the values of ϕ and ψ may be written

$$\phi = (\kappa/2\pi)\tan^{-1}y/x, \quad \psi = -(\kappa/2\pi)\log r.$$

99. The investigations of the last articles are kinematical, we shall now calculate the value of the pressure within and without the vortex.

Let the values of the quantities inside the vortex, be distinguished from those outside, by accented letters.

Outside the vortex
$$p/\rho = C - \phi - \tfrac{1}{2}q^2,$$
and since $\phi = 0$, and $q = m/\pi r = \kappa/2\pi r$, we obtain
$$p/\rho = C - \kappa^2/8\pi^2 r^2 \quad\quad\quad (11),$$
whence if Π be the pressure at an infinite distance
$$\frac{p}{\rho} = \frac{\Pi}{\rho} - \frac{\kappa^2}{8\pi^2 r^2} \quad\quad\quad (12).$$

The equation of motion inside the vortex is
$$\frac{1}{\rho}\frac{dp'}{dr} = \frac{q'^2}{r} = \frac{\kappa^2 r}{4\pi^2 a^4},$$
whence
$$\frac{p'}{\rho} = \frac{\kappa^2 r^2}{8\pi^2 a^4} + \frac{P}{\rho} \quad\quad\quad (13),$$
where P is the pressure at the centre of the vortex.

At the surface of the vortex where $r = a$, $p = p'$, whence
$$P/\rho = \Pi/\rho - \kappa^2/4\pi^2 a^2 \quad\quad\quad (14),$$
and therefore
$$\frac{p'}{\rho} = \frac{\Pi}{\rho} - \frac{\kappa^2}{4\pi^2 a^2}\left(1 - \frac{r^2}{2a^2}\right) \quad\quad\quad (15).$$

Hence if $\Pi < \kappa^2 \rho/4\pi^2 a^2$,

p' will become negative for some value of $r < a$, which shows that a cylindrical hollow will exist in the vortex, which is concentric with its outer boundary.

When there is no hollow, equations (12) and (15) show that the pressure is a minimum at the centre of the vortex, where it is equal to $\Pi - \kappa^2 \rho/4\pi^2 a^2$, and that it gradually increases until the surface is reached, at which it attains its maximum value, which is equal $\Pi - \kappa^2 \rho/8\pi^2 a^2$, and that it then diminishes to infinity, where its value is Π.

It is also possible to have a hollow cylindrical space, round which there is cyclic irrotational motion. Such a space is called a *hollow vortex*. The condition for its existence requires that $p = 0$, when $r = a$, and therefore by (12)
$$\Pi = \kappa^2 \rho/8\pi^2 a^2.$$

This equation determines the value of the radius of the hollow, when the pressure at a very great distance is given.

ELLIPTIC VORTEX.

100. Kirchhoff has shown that it is possible for a vortex whose cross section is an invariable ellipse, and whose molecular rotation at every point is constant, to rotate in a state of steady motion in an infinite liquid, provided a certain relation exists between the molecular rotation and the angular velocity of the axes of the cross section.

The current function is evidently equal to the potential of an elliptic cylinder of density $\zeta/2\pi$. Let a and b be the semi-axes of the cross section, then the value of ψ inside the vortex may be taken to be

$$\psi' = D - \zeta(Ax^2 + By^2)/(A + B),$$

where A, B, D are constants, for this value of ψ' satisfies (2).

Let $x = c\cosh\eta\cos\xi$, $y = c\sinh\eta\sin\xi$, where $c = (a^2 - b^2)^{\frac{1}{2}}$, and let $\eta = \beta$ at the surface; the value of ψ' becomes

$$\psi' = D - \zeta c^2(A\cosh^2\eta\cos^2\xi + B\sinh^2\eta\sin^2\xi)/(A + B).$$

Also let the value of ψ outside the vortex be

$$\psi = A'\epsilon^{-2\eta}\cos 2\xi + D\eta/\beta.$$

When $\eta = \beta$, we must have

$$\psi - \psi' = \text{const.}, \quad d\psi/d\eta = d\psi'/d\eta.$$

Therefore $\quad A'\epsilon^{-2\beta} = -\tfrac{1}{2}\zeta c^2(A\cosh^2\beta - B\sinh^2\beta)/(A+B)$

and $\quad\quad\quad A'\epsilon^{-2\beta} = \tfrac{1}{2}\zeta c^2(A-B)\sinh\beta\cosh\beta/(A+B).$

Whence $\quad A'(a-b)^2 = -\dfrac{\zeta c^2(Aa^2 - Bb^2)}{2(A+B)} = \dfrac{\zeta c^2(A-B)ab}{2(A+B)}.$

Therefore $Aa = Bb$ and

$$\psi' = D - \zeta(bx^2 + ay^2)/(a+b).$$

Let ω be the angular velocity of the axes; u, v the velocities of the liquid parallel to them, then

$$\dot{x} - y\omega = u = d\psi'/dy = -2a\zeta y/(a+b),$$
$$\dot{y} + x\omega = v = -d\psi'/dx = 2b\zeta x/(a+b).$$

The boundary condition is

$$\dot{x}\frac{dF}{dx} + \dot{y}\frac{dF}{dy} = 0,$$

where $F = (x/a)^2 + (y/b)^2 - 1 = 0$. Whence

$$\left(\omega - \frac{2a\zeta}{a+b}\right)\frac{1}{a^2} + \left(\frac{2b\zeta}{a+b} - \omega\right)\frac{1}{b^2} = 0,$$

therefore $\quad\quad\quad \omega = 2ab\zeta/(a+b)^2.$

We therefore obtain
$$\dot{x} = -a\omega y/b, \quad \dot{y} = b\omega x/a,$$
the integrals of which are
$$x = La\cos(\omega t + \alpha), \quad y = Lb\sin(\omega t + \alpha),$$
where L and α are the constants of integration. Whence the path of every particle relatively to the boundary, is a similar ellipse.

101. A complete investigation respecting the stability of a vortex, is given in my larger treatise; but it may be stated that when the cross section is circular, both cases of steady motion which we have considered, viz. the steady motion of a solid vortex, and the steady motion of a hollow vortex, are stable; and consequently if a small disturbance be communicated to either kind of vortex, the vortex will proceed to oscillate about its mean form in steady motion, and will not mix with the surrounding liquid. It is otherwise if the liquid composing the vortex is of different density to the surrounding liquid, for in that case the motion will be unstable; and consequently after a sufficient time has elapsed, the two liquids will have become mixed together, and will form what has been called a vortex sponge. The stability of Kirchhoff's elliptic vortex, does not appear to have been investigated.

The preceding results are however only applicable when there is a single vortex in an infinite liquid, and it is therefore important to enquire, whether the presence of other vortices or the presence of plane or curved boundaries renders the motion unstable. This question has been dealt with by Prof. J. J. Thomson, and he has shown that when there are two rectilinear vortices in a liquid, the linear dimensions of whose cross sections are small in comparison with the shortest distance between them, their cross sections will always remain approximately circular; and it is inferred from this that a similar result holds good in the case of any number of vortices. We therefore conclude that when a number of vortices of small cross section exist in a liquid, they may be treated as if their cross sections remained circular throughout the subsequent motion, provided none of the vortices approach too closely to one another. It therefore follows, that the effect of any number of vortices upon any external point of the liquid, is equal to the sum of the effects due to each; so that if $m_1, m_2, \ldots$

be the strengths of the vortices, r_1, r_2...... their distances from any point P of the liquid, the current function due to the whole motion is

$$\psi = -(m_1/\pi)\log r_1 - (m_2/\pi)\log r_2 - \ldots\ldots\ldots\ldots$$

Moreover since a rectilinear vortex is incapable of producing any motion of translation upon itself, it follows, that the motion of any particular vortex, is the same as would be produced by all the other vortices upon the point occupied by the particular vortex, if the latter did not exist.

102. We shall pass on to consider the motion of a number of vortices of small and approximately circular cross sections.

Putting $m/\pi = M$, it follows that since we neglect deformations of the cross sections, the current function due to each vortex will be $-M\log r$, and the velocity due to it at any point P will be M/r, and will be perpendicular to the line joining P with the vortex. Hence if two vortices of equal strengths exist in a liquid, each vortex will describe a circle whose centre is the middle point of the line joining them, with velocity $M/2c$, where $2c$ is the distance between them; and therefore each vortex will move as if there existed a stress in the nature of a tension between them, of magnitude $M^2/4c^3$.[1]

To find the stream lines relative to the line joining the vortices, take moving axes, in which the axis of x coincides with the above-mentioned line; then

$$\psi = -\tfrac{1}{2}M\log\{y^2 + (x-c)^2\}\{y^2 + (x+c)^2\}.$$

Also
$$\dot{x} - \omega y = u = d\psi/dy,$$
$$\dot{y} + \omega x = v = -d\psi/dx,$$

where $\omega = M/2c^2$. Let
$$\chi = \psi + \tfrac{1}{2}\omega(x^2 + y^2),$$

therefore $\dot{x} = d\chi/dy,\ \dot{y} = -d\chi/dx.$

Multiplying by $\dot{y}$, $\dot{x}$ respectively, subtracting and integrating, we obtain
$$\chi = \text{const.} = A,$$

whence the equation of the relative stream lines is

$$\tfrac{1}{2}\omega(x^2 + y^2) - \tfrac{1}{2}M\log\{y^2 + (x-c)^2\}\{y^2 + (x+c)^2\} = A.$$

[1] Greenhill, "Plane Vortex Motion," *Quart. Journ.* vol. xv. p. 20.

103. If two opposite vortices of strengths m and $-m$ are present in the liquid, the vortices will move perpendicularly to the line joining them with velocity $M/2c$, where $2c$ is the distance between them.

In this case there is evidently no flux across the plane which bisects the line joining the vortices, and which is perpendicular to it; we may therefore remove one of the vortices and substitute this plane for it. Hence a vortex in a liquid which is bounded by a fixed plane will move parallel to the plane, and the motion of the liquid will be the same as would be caused by the original vortex, together with another vortex of equal and opposite strength, which is at an equal distance and on the opposite side of the plane.

This vortex is evidently the image of the original vortex, and we may therefore apply the theory of images in considering the motion of vortices in a liquid bounded by planes.

104. If there is a vortex at the point (x, y) moving in a square corner bounded by the planes Ox, Oy, the images will consist of two negative vortices at the points $(-x, y)$, $(x, -y)$, and a positive vortex at the point $(-x, -y)$; for if these vortices be substituted for the planes, their combined effect will be to cause no flux across them.

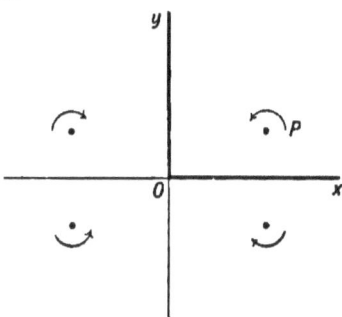

Since the vortex is incapable of producing any motion of translation upon itself, its motion will be due solely to that produced by the combined effect of its images; whence,

$$\dot{x} = \frac{M}{2y} - \frac{My}{2(x^2+y^2)} = \frac{Mx^2}{2y(x^2+y^2)},$$

$$\dot{y} = -\frac{M}{2x} + \frac{Mx}{2(x^2+y^2)} = -\frac{My^2}{2x(x^2+y^2)};$$

VORTEX IN A CIRCULAR CYLINDER.

therefore $\dot{x}/x^2 + \dot{y}/y^2 = 0$,

whence $x^{-2} + y^{-2} = a^{-2}$

or $r \sin 2\theta = 2a$.

This is the equation of a Cotes' Spiral, which is the curve described by the vortices: also since

$$x\dot{y} - \dot{x}y = -\tfrac{1}{2}M$$

the vortex describes the spiral in exactly the same way as a particle would describe it, if repelled from the origin with a force $3M^2/4r^3$.

105. The method of images may also be applied to determine the current function due to a vortex in a liquid, which is bounded externally or internally by a circular cylinder.

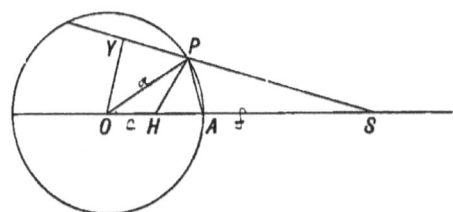

Let H be the vortex, a the radius of the cylinder, $OH = c$; and let S be a point such that $OS = f = a^2/c$, then the triangles SOP and POH are similar, therefore

$$SPO = OHP,$$
$$OPH = OSP,$$

also $\quad OSP + SPA = OAP = OPA$
$$= OPH + HPA,$$

therefore $\quad SPA = HPA$.

Let us place another vortex of equal and opposite strength at S, then the velocity along OP due to the two vortices is

$$u = -\frac{M}{HP} \sin HPO + \frac{M}{SP} \sin SPO.$$

But $\quad \dfrac{\sin HPO}{\sin SPO} = \dfrac{\sin HPO}{\sin OHP}$
$$= OH/a$$
$$= HP/SP,$$

hence $u = 0$ and there is no flux across the cylinder.

Hence the image of a vortex inside a cylinder, is another vortex of equal and opposite strength situated on the line joining the vortex with the centre of the cylinder, and at a distance a^2/c from the centre, and the vortex and its image will describe circles about the centre with a velocity

$$M/SH = Mc/(a^2 - c^2).$$

The velocities of the vortex and its image are equal, but their angular velocities about the axis of the cylinder will be different; hence the motion of the liquid inside the cylinder and the motion of the liquid outside the cylinder are independent, and the vortex and its image will not remain on the same radial plane in the subsequent motion. Hence the motions of the liquid inside and outside the cylinder do not correspond, as is the case with plane boundaries, except at the instant when the vortex and its image are on the same radial plane.

The current function of the liquid at a point (r, θ) within the cylinder is

$$\psi = -M \log HP/SP$$
$$= -\tfrac{1}{2}M \log \frac{r^2 + c^2 - 2rc \cos \theta}{r^2 + f^2 - 2rf \cos \theta}.$$

106. We have shown that the velocity potential due to a source is $m \log r$; hence if we have a combination of a source of strength m, and a vortex of strength m', the velocity potential due to the two is

$$\phi = m \log r + M \tan^{-1} y/x,$$

where $M = m'/\pi$. Whence

$$u = \frac{mx - My}{x^2 + y^2}, \quad v = \frac{my + Mx}{x^2 + y^2}.$$

An arrangement of this kind is called *Rankine's free spiral vortex*.

In order to find the stream lines, let us transfer to polar co-ordinates, and we find

$$\frac{dr}{dt} = \frac{m}{r}, \quad r\frac{d\theta}{dt} = \frac{M}{r},$$

whence if $m/M = \alpha$, we obtain

$$r = A\epsilon^{\alpha\theta},$$

and therefore the stream lines are equiangular spirals.

107. We shall conclude this Chapter by proving three fundamental properties of vortex motion.

We have defined a vortex line to be a line whose direction coincides with the direction of the instantaneous axis of molecular rotation. If through every point of a small closed curve a series of vortex lines be drawn, they will enclose a volume of fluid which may be called a vortex filament, or shortly a vortex.

We have shown that if the forces which act on the fluid have a potential, and the density is a function of the pressure, the motion of the fluid constituting the vortex can never become irrotational. It will now be shown that every vortex possesses the following three fundamental properties:

(i) *Every vortex is always composed of the same elements of fluid.*

(ii) *The product of the angular velocity of any vortex into its cross section, is constant with respect to the time, and is the same throughout its length.*

(iii) *Every vortex must either form a closed curve, or have its extremities in the boundaries of the fluid.*

To prove the first proposition, let P and Q be any two adjacent points on a vortex, ω the molecular rotation at P. Then by the definition of a vortex line, PQ is the direction about which the rotation ω takes place.

Let P', Q' be the positions of P and Q at the end of an interval δt; then we have to show that $P'Q'$ is the instantaneous axis of rotation at P'.

Let x, y, z be the coordinates of P; u, v, w the velocities of the element of fluid which at time t is situated at P.

If $PQ = h$, the coordinates of Q are evidently
$$x + h\xi/\omega, \quad y + h\eta/\omega, \quad z + h\zeta/\omega;$$
also since $u = F(x, y, z, t)$, it follows that if u_1, v_1, w_1 be the velocities of Q,
$$u_1 = F(x + h\xi/\omega, y + h\eta/\omega, z + h\zeta/\omega, t)$$
$$= u + \frac{h}{\omega}\left(\xi\frac{du}{dx} + \eta\frac{du}{dy} + \zeta\frac{du}{dz}\right),$$
$$= u + \frac{h\rho}{\omega}\frac{\partial}{\partial t}\left(\frac{\xi}{\rho}\right) \quad \ldots\ldots\ldots\ldots\ldots \quad \ldots\ldots\ldots\ldots\ldots (16).$$

by § 18.

The coordinates of P' are

$$x + u\delta t,\ y + v\delta t,\ z + w\delta t,$$

and those of Q' are

$$x + h\xi/\omega + u_1\delta t = x + u\delta t + \frac{h\rho}{\omega}\left\{\frac{\xi}{\rho} + \frac{\partial}{\partial t}\left(\frac{\xi}{\rho}\right)\delta t\right\}$$

$$= x + u\delta t + \frac{h\rho\xi'}{\omega\rho'},$$

by (16), where ρ' is the density, and ξ', η', ζ', the components of molecular rotation at P'.

Hence if h' denote the length of $P'Q'$, and λ', μ', ν' its direction cosines, then

$$\lambda'h' = h\rho\xi'/\omega\rho',\quad \mu'h' = h\rho\eta'/\omega\rho',\quad \nu'h' = h\rho\zeta'/\omega\rho'\ \ldots\ldots(17),$$

whence
$$\lambda'/\xi' = \mu'/\eta' = \nu'/\zeta',$$

which shows that $P'Q'$ is the instantaneous axis of rotation at P', and therefore $P'Q'$ is the element of the vortex line, which at time t occupied the position PQ. This proves the first theorem.

To prove the second theorem, square and add (17) and we obtain

$$h' = h\rho\omega'/\omega\rho'.$$

But since the mass of the element is constant

$$\rho h\sigma = \rho'h'\sigma',$$

whence
$$\sigma\omega = \sigma'\omega',$$

which proves that $\sigma\omega$ is independent of the time.

Since
$$\frac{d\xi}{dx} + \frac{d\eta}{dy} + \frac{d\zeta}{dz} = 0,$$

it can be shown by integrating this expression by parts, throughout the interior of any closed surface, as was done in proving Green's Theorem, that

$$\iint (l\xi + m\eta + n\zeta)\, dS = \iiint \left(\frac{d\xi}{dx} + \frac{d\eta}{dy} + \frac{d\zeta}{dz}\right) dxdydz = 0,$$

or $\iint \omega \cos \epsilon\, dS = 0,$

where ϵ is the angle between the axis of rotation and the normal to S drawn outwards.

Now if we choose S so as to coincide with the surface of any finite portion of a vortex of small section, together with its two ends, cos ε vanishes except at the two ends; and is equal to $+1$ at one end, and -1 at the other; hence

$$\omega_1 dS_1 - \omega_2 dS_2 = 0,$$

which proves the second part of (ii).

To prove the third theorem, we observe that if a vortex did not form a closed curve or have its extremities in the boundary, it would be possible to draw a closed surface cutting the vortex once only, and the surface integral would not vanish.

The first theorem and the first part of the second theorem depend on dynamical considerations; the second part of this theorem and the third theorem are kinematical.

EXAMPLES.

1. If the axis of a hollow vortex be the axis of z, measured vertically downwards, the plane of xy being the asymptotic plane to the free surface, and if ϖ be the atmospheric pressure: prove that the equation of the surface at which the pressure is $\varpi + g\rho a$ is

$$(x^2 + y^2)(z - a) = c^3,$$

where c is a constant.

2. Three rectilinear vortices of equal strengths form the edges of an equilateral triangular prism. Prove that they will always form the three edges of an equal prism.

3. If n rectilinear vortex filaments of equal strengths, be initially at the angles of a prism whose base is a regular polygon of n sides, show that they will always be so situated, and that each filament will describe the circumscribed cylinder with velocity $k(n-1)/2a$ where k is the velocity due to each vortex at unit distance and a is the radius of the cylinder. Show also that the equation of the relative stream lines referred to the radius through a vortex as initial line is $r^{2n} - 2a^n r^n \cos n\theta - b^{2n} = 0$.

4. The space on one side of the concave branch of a rectangular hyperbolic cylinder is filled with liquid, and a rectilinear vortex exists in the liquid; prove that it moves in a cylinder having the same asymptotic planes as the boundary.

5. The motion of a liquid in two dimensions is such that the rotation ζ is constant; prove that the general functional equation of the stream lines is

$$\phi(y + \iota x) + \chi(y - \iota x) - \tfrac{1}{2}\zeta(x^2 + y^2) = c.$$

Prove that if the space between one branch of the hyperbola $x^2 - 3y^2 = a^2$ and the tangent to its vertex be filled with liquid, it will be possible for the liquid to move steadily with constant rotation, and find the form of the stream lines.

6. A mass of liquid whose outer boundary is an infinitely long cylinder of radius b, is in a state of cyclic irrotational motion and is under the action of a uniform pressure Π over its external surface. Prove that there must be a concentric cylindrical hollow whose radius a is determined by the equation

$$8\pi^2 a^2 b^2 \Pi = M\kappa^2,$$

where M is the mass of unit length of the liquid, and κ is the circulation.

If the cylinder receive a small symmetrical displacement, prove that the time of a small oscillation is

$$\frac{4}{\kappa}\pi^2 a^2 b^2 \sqrt{\frac{\log b/a}{b^4 - a^4}}.$$

7. Four straight vortex filaments with alternately positive and negative rotations are placed symmetrically within a cylinder filled with liquid; prove that if the motion is steady, the distance of each filament from the axis of the cylinder is nearly three-fifths of the radius of the latter.

8. Prove that three infinitely long straight cylindrical vortices of equal strengths will be in stable steady motion, when situated at the vertices of an equilateral triangle whose sides are large compared with the radii of the sections of the vortices; and that if they are slightly displaced, prove that the time of a small oscillation is the same as that of the time of revolution of the system in its undisturbed state.

9. A straight cylindrical vortex column of uniform rotation ζ, is surrounded by an infinite quantity of liquid moving irrotationally which is at rest at infinity; prove that the difference between the kinetic energy included between two planes at right angles to the axis of the cylinder and separated by unit distance

when the cross section is an ellipse, and when it is a circle of equal area A is
$$\rho\pi^{-1}\zeta^2 A^2 \log(a+b)/2\sqrt{(ab)},$$
where ρ is the density of the liquid, and a and b are the semiaxes of the ellipse.

10. A quantity of liquid whose rotation is uniform and equal to ζ, and whose external surface is a circular cylinder, surrounds a concentric cylinder of radius a. The external surface is subjected to a constant pressure Π. Prove that if the inner cylinder be removed, the velocity of the internal surface when its radius is α is equal to
$$\frac{1}{\alpha}\sqrt{\frac{(a^2-\alpha^2)(\zeta^2 c^2 - 2\Pi/\rho)}{\log \alpha^2/(\alpha^2+c^2)}},$$
where $\pi\rho c^2$ is the mass of the liquid per unit of length.

11. If a vortex is moving in a liquid bounded by a fixed plane, prove that a stream line can never coincide with a line of constant pressure.

12. If a pair of equal and opposite vortices are situated inside or outside a circular cylinder of radius a, prove that the equation of the curve described by each vortex is,
$$(r^2-a^2)^2(r^2\sin^2\theta - b^2) = 4a^2 b^2 r^2 \sin^2\theta,$$
where b is a constant.

PART II.

THEORY OF SOUND.

CHAPTER VI.

INTRODUCTION[1].

108. SOUNDS may be divided into two classes, musical sounds or notes, and unmusical sounds or noises, and a little consideration is sufficient to show that the former is the simpler phenomenon of the two. If, for example, one of the keys of a pianoforte be struck, we obtain a musical note, whereas if all the notes are sounded together, the result is an unmusical sound or noise, in which the different notes cannot be distinguished. We thus see that a combination of a number of musical notes will produce a noise, but on the other hand no combination of noises is capable of producing a musical note.

The sensation of sound is produced by means of vibrations of the atmosphere, which are first communicated to the tympanum of the ear, and are afterwards transmitted by the auditory nerve to the brain. That sound is produced by aerial vibrations, can be experimentally verified in a number of ways. Thus if a bell be placed under the receiver of an air pump, and the air is gradually exhausted, the sound produced by the bell becomes fainter and fainter, and at last ceases to be heard. If again, a note is produced by striking an ordinary finger bowl, the latter will be thrown into a state of tremor or vibration, the existence of which can be perceived by cautiously touching the bowl with the fingers; and if the vibrations are stopped by pressing the bowl between the hands, the note ceases upon the stoppage of the vibrations. In this case, the vibrations of the bowl are communicated to the atmosphere, and waves are propagated through the latter in all directions from the bowl.

[1] This and the following Chapter have been principally derived from vol. I. of Lord Rayleigh's Treatise.

109. At the commencement of Chapter IV. we explained the kinematics of wave motion, and we showed that the properties of a wave depend upon three quantities, viz. the *amplitude*, the *velocity of propagation* V, and the *wave length* λ. We may also, if we please, introduce the period τ instead of the wave length, since $V\tau = \lambda$. Notes also have three characteristics, viz. *intensity*, *pitch*, and a quality called *timbre*; and we must now enquire how these three physical characteristics of a note, are connected with the geometrical constants of a wave.

110. In the first place it can be shown that the velocity of all notes in air and gases is very nearly the same[1]; for if this were not the case, a piece of music which is played in tune, would become hopelessly discordant when heard by an observer situated at a considerable distance. A similar proposition is true of all substances which are capable of propagating sound; although the magnitude of the velocity of propagation depends upon the particular substance, being greater in the case of solids and liquids than in gases. Thus in dry air, the velocity of sound at $0°$ C. is about 332 metres (i.e. 1089 feet) per second, whilst in water at $8°$ C. it is about 1435 metres (i.e. 4708 feet) per second[2]. It therefore follows that the properties of a note do not depend upon its velocity of propagation.

111. The intensity of sound is measured by the rate at which energy is propagated across a given area parallel to the waves, and is proportional to the square of the amplitude.

112. The pitch of a note is the quality by which its place in the musical scale is recognised. Thus the middle c of a pianoforte, is said to have a pitch an octave lower than the next succeeding c in the scale. We shall now show that the pitch of a note depends upon the *frequency* of vibration, which has in Chapter IV. been defined to be the number of vibrations executed per second; and that the pitch rises as the frequency increases.

This is most easily shown by means of an apparatus called the

---

[1] It appears however, that violent sounds, such as are caused by explosions, travel with a higher velocity than sounds produced by notes. Experiments made by Krupp's firm at Essen, for the purpose of ascertaining the velocity with which the reports of heavy guns travel, showed that the velocity may amount to 2034 feet per second. See "Sound velocity applied to range finding," Captain G. G. Aston, *Proc. Roy. Artillery Inst.* April, 1890.

[2] To reduce metres to feet, multiply by 3·2809.

Siren, invented by Cagniard de la Tour. This instrument consists of a stiff circular disc, which is capable of revolving about an axis, and is pierced with one or more sets of holes arranged at equal intervals around its circumference. A wind pipe in connection with bellows is presented perpendicularly to one of the holes, and the disc is made to revolve. When the time of revolution is small, the wind escapes by means of a succession of puffs; but after the time of revolution has sufficiently increased, the puffs blend into a single note of definite pitch; and if the time of revolution is still further increased, the pitch of the note rises in the scale. *This shows that the pitch of a note depends upon the frequency.* Another point of importance is, that if the time of revolution is doubled, the two notes stand to one another in the relation of octaves; so that if f be the frequency of any particular note, the frequency of the note an octave higher is $2f$.

If $\epsilon^{\iota n t}$ be the time factor of a vibration, the frequency is $n/2\pi$; but since $n = 2\pi V/\lambda$, the frequency is also equal to V/λ; we have also shown that the pitch is independent of V, and it therefore follows that the frequency varies inversely as the wave length; consequently the shorter the wave length, the higher the pitch of the note.

The frequency of a given note, is to a slight extent arbitrary, in as much as the ear is incapable of distinguishing slight differences of pitch. At the Stuttgart conference in 1834, it was recommended that the middle c of a pianoforte, which is written c', should correspond to 264 vibrations per second. The pitch usually adopted by acoustical instrument makers, is taken to be $c' = 256$ or 2^8 vibrations per second, so that the frequencies of the octaves and sub-octaves are represented by powers of 2. Hence the wave length of c' is about 4·2 feet.

Trained ears are capable of recognising an enormous number of gradations of pitch, but in as much as the power of perception varies with different ears, it is somewhat difficult to assign limits to the audibility of notes. It is probable that the perception of pitch begins[1] when the number of vibrations in a second lies between 8 and 32, and ceases before it amounts to 40000.

113. All notes which are produced by musical instruments are of a highly compound nature; and when we discuss the

---

[1] Donkin's *Acoustics*, § 19.

dynamical part of the subject, it will be shown that the vibrations which are capable of being produced by a vibrating body, are usually represented by an infinite series of terms of the form Ae^{int}, in which the frequency of each term is different. A note which the ear is incapable of resolving, is called by Helmholtz a *tone*. We thus see that a note which is represented by a series of terms of the type Ae^{int}, is a compound note consisting of a number of tones, which are represented by the different terms of the series. The component tone of this series, whose frequency is the least, is called the *gravest* or *fundamental* tone, and the other tones are called overtones. It frequently happens (although there are exceptions), that the amplitudes of the component tones diminish as their pitches rise, so that the amplitude of the gravest tone is sufficiently large to impress its character upon the whole vibration; and in many cases is the note which is most distinctly heard. Lord Rayleigh states[1], that he has recently examined a large metal bell weighing about 3 cwt., and that the following tones could be plainly heard[2], viz.

$$e\flat, \quad f'^{\sharp}_{\sharp}, \quad e'', \quad b''.$$

The gravest tone $e\flat$ had a long duration. When the bell was struck by a *hard* body, the higher tones were at first predominant, but after a time they died away leaving $e\flat$ in possession of the field. When the striking body was soft, the original preponderance of the higher elements was less marked.

114. The word *timbre* is used to express a quality by which notes of the same intensity and pitch are distinguishable from one another, and which probably depends upon the nature of the instrument employed in producing the note.

115. Another phenomenon which we must notice, is that of *beats*. Let us suppose for simplicity that two notes of the same amplitude and phase, have slightly different frequencies m and n. The vibration produced by the combination of these two notes, may be represented by the equation

$$y = a \cos 2\pi mt + a \cos 2\pi nt$$
$$= 2 \cos \pi (m - n) t \cos \pi (m + n) t$$
$$= 2 \cos \pi (m - n) t \cos \{2\pi m - \pi (m - n)\} t.$$

[1] On Bells, *Phil. Mag.* Jan. 1890.
[2] c is the octave below, and c'' is the octave above the middle c of the pianoforte.

BEATS.

Since $m - n$ is small, the resultant vibration may be regarded as one, whose amplitude and phase vary slowly with the time. We thus see that the amplitude vanishes whenever

$$t = (2s + 1)/2\,(m - n),$$

and is a maximum when $t = s/(m - n)$, where s is zero or any positive integer. Hence at intervals $(m - n)^{-1}$ there is absolute silence, and midway between the intervals of absolute silence, the intensity of the sound attains its maximum value. The intervals of silence are called *beats*, and the number of beats per second is $m - n$.

In order that beats may be heard distinctly, $m - n$, or the difference between the frequencies of the two notes, must be small.

CHAPTER VII.

VIBRATIONS OF STRINGS AND MEMBRANES.

116. IF a piece of string or wire be tightly stretched between two fixed points, and be set in motion, either by being struck or rubbed with a bow, it is well known that a musical note will be produced. This arises from the circumstance that the string or wire is set into vibration, and we shall now proceed to investigate the theory of these vibrations.

If a thin metal wire, whose natural form is straight, is bent into a plane curve of any form, the resultant stresses across any normal section, due to the action of contiguous portions of the wire, consist of (i) a tension T perpendicular to the section, (ii) a normal shearing stress N, (iii) a flexural couple G, whose axis is perpendicular to the plane of the wire; and consequently in order to investigate the vibrations of instruments whose strings are made of wire, it would be necessary to construct a theory which would take all three stresses into account. It is however obvious, that although a thin string, made for example of catgut, is capable of sustaining a considerable tension, the resistance which it is capable of offering to shearing stress and to bending, is very small in comparison with the resistance which it is capable of offering to stretching. We may therefore when dealing with strings made of catgut and similar materials, neglect the shearing stress and the couple, and may treat the string as *perfectly flexible*. We may define a perfectly flexible string to be a string, which is incapable of offering any resistance to shearing stress or to bending. A string of this kind is an ideal substance which does not exist in nature; but in as much as most thin strings which are not made of wire, approximate to the condition of perfect flexibility, it will be desirable first of all to consider the vibrations of a perfectly flexible string. If however the string is too stiff to be treated as perfectly flexible, or is made of wire, the theory of the vibrations which it is capable of executing, fall more properly under the head of the vibrations of bars. These will be considered in the next chapter.

The vibrations which a stretched string is capable of executing, consist of two kinds, which may be treated as being independent of one another. The first kind consists of *transverse* vibrations, in which the displacement of every element is *perpendicular* (or very approximately so), to the undisplaced position of the string. The second class consists of *longitudinal* vibrations, in which the displacement is *parallel* to the undisplaced position of the string. It will thus be seen, that in the theory of transverse vibrations, the longitudinal displacement is supposed to be so small in comparison with the transverse displacement, that the former may be neglected in comparison with the latter.

Transverse Vibrations of Strings.

117. To find the equation of motion for transverse vibrations, it will be sufficient to consider the case in which the motion takes place in a plane. Let T_1 be the tension, ρ the linear density, i.e. the mass of a unit of length, y the displacement of the point whose abscissa is x, Y the impressed force per unit of mass.

If ϕ be the angle which any element δs makes with the axis of x, the equation of motion is

$$\rho \ddot{y} \delta s = \frac{d}{ds}(T_1 \sin \phi)\, \delta s + \rho Y \delta s.$$

Now $\sin \phi = dy/ds$; also since the displacement y is small, the curvature will also be small, and we may therefore put $ds = dx$. The tension T_1 may also be regarded as constant throughout the length of the string, whence the equation of motion becomes

$$\frac{d^2 y}{dt^2} = \frac{T_1}{\rho} \frac{d^2 y}{dx^2} + Y \quad \ldots\ldots\ldots\ldots\ldots\ldots\ldots(1).$$

If the motion does not take place in a plane, we may resolve the displacements and forces into two components respectively parallel to the axes of y and z, and we shall thus obtain a second equation of the same form as (1), in which z, Z are written for y, Y respectively.

118. Let us now suppose that the length of the string is equal to l, and that there are no impressed forces; also let

$$a^2 = T_1/\rho \quad \ldots\ldots(2).$$

Equation (1) now becomes

$$\frac{d^2y}{dt^2} = a^2 \frac{d^2y}{dx^2} \quad \ldots\ldots(3).$$

To solve this equation, assume

$$y = F(x)\, \epsilon^{\imath mat}.$$

Substituting in (3) we obtain

$$\frac{d^2F}{dx^2} + m^2 F = 0,$$

the solution of which is

$$F = C \sin mx + D \cos mx.$$

The solution of (3) may therefore be written in the form

$$y = \Sigma\, (C \sin mx + D \cos mx)\, \epsilon^{\imath mat} \ldots\ldots(4),$$

where m is at present undetermined, and C and D are complex constants.

The value of m will depend upon the particular problem under consideration. We shall now suppose that both ends of the string are fixed; in this case the conditions to be satisfied at the fixed ends are, that y and $\dot{y}$ should vanish when $x = 0$ and $x = l$. These conditions evidently require that

$$D = 0, \quad \sin ml = 0;$$

from the last of which we deduce,

$$m = s\pi/l,$$

where s is a positive integer. Writing $C = A - \imath B$ and rejecting the imaginary part, the solution becomes

$$y = \Sigma_1^\infty (A_s \cos s\pi at/l + B_s \sin s\pi at/l) \sin s\pi x/l \ \ \ldots\ldots(5),$$

and therefore the period τ_s of the sth component, is given by

$$\tau_s = \frac{2l}{sa} = \frac{2l}{s}\sqrt{\frac{\rho}{T_1}},$$

and the frequency

$$f_s = \frac{s}{2l}\sqrt{\frac{T_1}{\rho}}.$$

FREQUENCY OF VIBRATION. 133

The gravest note corresponds to $s = 1$, and therefore its frequency is

$$f_1 = \frac{1}{2l}\sqrt{\frac{T_1}{\rho}}.$$

From these results we draw the following conclusions.

(i) The frequency is inversely proportional to the length; and therefore if the string be shortened, the pitch of the note will rise, and conversely if the string be lengthened, the pitch will fall. We thus see why it is that in playing a violin, different notes can be obtained from the same string.

(ii) The frequency is proportional to the square root of the tension, accordingly if the string be tightened the pitch will rise.

(iii) The frequency is inversely proportional to the square root of the density; and therefore if two strings having the same lengths, cross sections and tensions, be made of catgut and metal respectively, the pitch of the note yielded by the catgut string, will be higher than that yielded by the metal string; also the pitch of the note yielded by a thick string, will be graver than that of the note yielded by a thin string, of the same material, length and tension.

If s be any integer other than unity, we learn from (5) that the displacement is zero at all points for which $x = rl/s$, where $r = 1, 2, 3, \ldots s - 1$; it therefore follows, that corresponding to the sth harmonic, there are $s - 1$ points situated at equal intervals along the string, at which there is no motion. These points are called *nodes*.

119. The constants A and B depend upon the initial circumstances of the motion. Now the motion of dynamical systems of which a string is an example, may be produced either by displacing every point in any arbitrary manner, subject to the condition that the connections of the system are not violated; or by imparting to every point an arbitrary initial velocity, subject to the same condition. Hence the most general possible motion, is obtained by communicating to every point of the string an initial displacement, and an initial velocity. We shall now show that when the initial displacements and velocities are given, the constants A and B are completely determined.

Let y_0, $\dot{y}_0$ be the initial displacements and velocities. Then it follows from (5) that

$$y_0 = \Sigma_1^\infty A_s \sin s\pi x/l \quad\dotfill(6),$$

$$\dot{y}_0 = \Sigma_1^\infty (s\pi a/l) B_s \sin s\pi x/l \quad\dotfill(7).$$

Now the integral $\int_0^l \sin(s\pi x/l) \sin(s'\pi x/l)\,dx$ is equal to zero if s and s' are different integers, and is equal to $\tfrac{1}{2}l$ if $s = s'$; whence multiplying (6) by $\sin s\pi x/l$ and integrating between the limits l and 0, we obtain

$$A_s = \frac{2}{l} \int_0^l y_0 \sin \frac{s\pi x}{l}\,dx \quad\dotfill(8).$$

Similarly from (7)

$$B_s = \frac{2}{s\pi a} \int_0^l \dot{y}_0 \sin \frac{s\pi x}{l}\,dx \quad\dotfill(9).$$

Since y_0, $\dot{y}_0$ are given functions of x, these equations completely determine the constants. We notice that B_s is zero when the initial velocity is zero, and that A_s is zero when there is no initial displacement.

120. As an example of these formulae, let us suppose that a point P, whose abscissa is b, of a string fixed at A and B, is displaced to a distance γ and then let go.

From $x = 0$ to $x = b$, $y_0 = \gamma x/b$, and therefore for this portion of the string

$$A_s' = \frac{2\gamma}{bl} \int_0^b x \sin \frac{s\pi x}{l}\,dx = \frac{2\gamma}{b}\left(-\frac{b}{s\pi}\cos\frac{s\pi b}{l} + \frac{l}{s^2\pi^2}\sin\frac{s\pi b}{l}\right).$$

From $x = b$ to $x = l$, $y_0 = \gamma(l-x)/(l-b)$, whence

$$A_s'' = \frac{2\gamma}{l(l-b)} \int_b^l (l-x)\sin\frac{s\pi x}{l}\,dx$$

$$= \frac{2\gamma}{b}\left\{\frac{b}{s\pi}\cos\frac{s\pi b}{l} + \frac{lb}{(l-b)s^2\pi^2}\sin\frac{s\pi b}{l}\right\}.$$

Whence adding we obtain

$$A_s' + A_s'' = A_s = \frac{2\gamma l^2}{s^2\pi^2 b(l-b)}\sin\frac{s\pi b}{l},$$

which determines A_s. This result shows that the amplitude of the gravest tone, which corresponds to $s = 1$, is greater than the amplitudes of any of the overtones. The gravest tone is therefore the most predominant.

121. The vibrations of a string which is set in motion by means of an initial velocity communicated *to every point of it*, may be investigated in a similar manner; but in many practical applications, a string is set in motion by means of an impulse *applied at some particular point*. The reader who desires to study the theory of the vibrations of a pianoforte wire, or of a violin string, is recommended to consult Donkin's *Acoustics*, and Lord Rayleigh's *Theory of Sound*. We shall confine ourselves to the simple case of the motion produced by an impulse F, applied at the point $x = b$.

Puting $n = s\pi a/l$, the value of y will be

$$y = \Sigma B_s \sin(nx/a) \sin nt \quad \ldots\ldots\ldots\ldots(10),$$

in which B_s has to be determined.

Now the work done by an impulse, is equal to half the product of the impulse and the initial velocity of the point at which it is applied; and since the work done is equal to the kinetic energy of the initial motion, we immediately obtain the equation

$$F\dot{\eta}_0 = \rho \int_0^l \dot{y}_0^2 dx \quad \ldots\ldots\ldots\ldots\ldots(11),$$

where η is the value of y at the point at which the impulse is applied.

From (10), it follows that

$$F\dot{\eta}_0 = F\Sigma B_s n \sin nb/a,$$

and

$$\rho \int_0^l \dot{y}_0^2 dx = \rho \int_0^l \Sigma (B_s n \sin nx/a)^2 dx$$

$$= \tfrac{1}{2} \rho l \Sigma B_s^2 n^2;$$

and therefore restoring the value of n, (11) becomes

$$F\Sigma B_s s \sin s\pi b/l = \tfrac{1}{2}\rho\pi a \Sigma B_s^2 s^2 \quad \ldots\ldots\ldots(12).$$

Comparing both sides of this equation, we see that

$$B_s = \frac{2F}{\pi\rho a s} \sin \frac{s\pi b}{l},$$

and therefore

$$y = \frac{2F}{\pi\rho a} \Sigma \frac{1}{s} \sin \frac{s\pi b}{l} \sin \frac{s\pi x}{l} \sin \frac{s\pi a t}{l}.$$

At the nodes corresponding to the sth component, we have $x = rl/s$; and since in the preceding expression the sth component of the displacement vanishes when $b = rl/s$, it follows that when the impulse is applied at a node, the corresponding component is absent.

122. We must now consider the motion of a string which is under the action of a periodic force $F(x) \cos pt$. It is well known that when an elastic body is set into vibration and left to itself, the motion gradually dies away and the system ultimately comes to rest. The reason of this is, that all such systems possess a property called viscosity or internal friction, by virtue of which the kinetic energy of the motion is gradually converted into heat. The effect of internal friction may be represented mathematically, by supposing that every element is retarded by a force proportional to its velocity, and we shall find it convenient in discussing motion due to a periodic force, to include the effect of viscosity.

It therefore follows that in (1) we must put
$$Y = F(x) \cos pt - k\dot{y},$$
where k is a constant, and the equation becomes
$$\frac{d^2y}{dt^2} + k\frac{dy}{dt} = a^2 \frac{d^2y}{dx^2} + F(x) \cos pt \ldots\ldots\ldots\ldots(13).$$

Since the motion which we are considering is periodic with respect to x as well as to t, we may assume
$$y = ue^{\iota mx}, \quad F(x) = E e^{\iota mx},$$
where $m = s\pi/l$; whence if $m^2 a^2 = n^2$, (13) becomes
$$\frac{d^2u}{dt^2} + k\frac{du}{dt} + n^2 u = E \cos pt \ldots\ldots\ldots\ldots(14).$$

The solution of this equation consists of two parts, viz. any particular solution of (14), together with the complementary function, which is the solution of the equation obtained by putting the right-hand side equal to zero, and which therefore contains two arbitrary constants. To find a particular solution, let us assume
$$u = c \cos(pt - e).$$

Substituting in (14), we obtain
$$c(n^2 - p^2) \cos(pt - e) - kpc \sin(pt - e) = E \cos e \cos(pt - e)$$
$$- E \sin e \sin(pt - e);$$
whence equating coefficients of $\sin(pt-e)$, $\cos(pt-e)$, we obtain
$$c(n^2 - p^2) = E \cos e$$
$$cpk = E \sin e,$$

FORCED AND FREE VIBRATIONS. 137

whence
$$u = \frac{E \sin e}{pk} \cos(pt - e) \ldots\ldots\ldots\ldots\ldots(15),$$

$$\tan e = \frac{pk}{n^2 - p^2} \ldots\ldots\ldots\ldots\ldots\ldots\ldots(16).$$

In order to obtain the complementary function, let $u = \epsilon^{qt}$; substituting in (14) and putting $E = 0$, we obtain

$$q^2 + kq + n^2 = 0,$$

the roots of which are

$$q = -\tfrac{1}{2}k \pm \iota \sqrt{(n^2 - \tfrac{1}{4}k^2)}.$$

Since k is always a small quantity, n will usually be greater than $\tfrac{1}{2}k$; it is moreover obvious that if n were less than $\tfrac{1}{2}k$, the time factor would be of the form $\epsilon^{(-\alpha \pm \beta)t}$, and would therefore not be periodic with respect to t. The complete solution of (14) may therefore be written

$$u = A\epsilon^{-\tfrac{1}{2}kt} \cos\{\sqrt{(n^2 - \tfrac{1}{4}k^2)}\,t - \alpha\} + \frac{E \sin e}{pk} \cos(pt - e)\ldots(17).$$

123. The first term of this equation represents the *free* vibrations; that is to say the vibrations which the string is capable of executing, when it is set in motion in any manner and then left to itself. The period of these vibrations is $2\pi(n^2 - \tfrac{1}{4}k^2)^{-\tfrac{1}{2}}$, and the amplitude is proportional to $\epsilon^{-\tfrac{1}{2}kt}$; the free vibrations therefore diminish as the time increases and ultimately die away.

In dissipative systems, it is usual to express the effect of friction by means of a quantity called the *modulus of decay*; which is defined to be the time which must elapse before the amplitude has fallen to ϵ^{-1} of its original value. It therefore follows that if τ be the modulus of decay,

$$\tau = 2/k,$$

which determines the physical meaning of k. We thus see that if the friction is small, so that the amplitude diminishes very slowly with the time, τ must be large, and therefore k must be small.

In order to pass to the case of no friction, we must put $k = 0$, in which case the frequency is proportional to n. Hence one of the effects of friction is to lower the pitch of the free vibrations.

Since the free vibrations always disappear after a sufficient time has elapsed, the expression for u ultimately reduces to the last term, which represents the forced vibrations, and which we shall proceed to consider.

124. A *forced vibration*, is a vibration produced and maintained by an external force. Its period is the same as that of the force, and it is consequently independent of the dimensions or constitution of the system. The amplitude of the forced vibration in the present case, when expressed in terms of n, p and k is

$$\frac{E}{\{(n^2 - p^2)^2 + p^2k^2\}^{\frac{1}{2}}}.$$

If the system were absolutely devoid of friction, k would be zero, and the amplitude would be infinite when $n = p$. In practical applications k is usually small; and we thus obtain the important theorem, that *if a system is acted upon by a periodic force, whose period is equal, or nearly so, to one of the periods of the free vibrations of the system, the corresponding forced vibration will be large.*

This theorem can be illustrated in the case of stringed instruments; for if a note be sounded whose period is the same, or nearly the same, as that of the fundamental note of one of the strings, the string will often be heard to vibrate in unison with the note; whereas if the period of the note be different from that of any of the natural periods of the string, no sound will be heard.

125. When both extremities of the string are fixed, the general solution of (13) which of course includes (3) as a particular case, may be presented in the following form, which is frequently useful.

In this case $m = s\pi/l$, and therefore $n = s\pi a/l$, whence the complete solution may be written

$$y = \Sigma_1^\infty \phi_s \sin s\pi x/l \ldots\ldots\ldots\ldots(18),$$

where ϕ_s is a function of the time which satisfies (14), and whose value is therefore determined by (17). The quantities denoted by ϕ_s, are called *normal functions*; and we shall now prove that the expressions for the kinetic and potential energies, do not contain any of the products of the normal functions. This is the characteristic property of these functions.

If T be the kinetic energy, we have

$$T = \tfrac{1}{2}\rho \int_0^l \dot{y}^2 dx = \tfrac{1}{2}\rho \int_0^l \{\Sigma_1^\infty \dot\phi_s \sin s\pi x/l\}^2 dx.$$

Since all the products vanish when integrated between the limits, we obtain

$$T = \tfrac{1}{4}\rho l \Sigma_1^\infty \dot{\phi}_s^2 \qquad \ldots\ldots\ldots(19).$$

The potential energy is equal to the work done in displacing the string to its actual position. In order to calculate its value, let the string be held in equilibrium in its actual configuration at time t by means of a force Y applied at every point of its length. The value of this force per unit of mass is equal to

$$-\frac{T_1}{\rho}\frac{d^2y}{dx^2},$$

by (1). Let δV be the work which must be done by this force in order to displace every element of the string through a space δy; then the work done upon an element δs

$$Y\rho \delta s \delta y = - T_1 \frac{d^2y}{dx^2} \delta s \delta y,$$

and therefore since $\delta s = \delta x$, the whole work done is

$$\delta V = - T_1 \int_0^l \frac{d^2y}{dx^2} \delta y\, dx.$$

Integrating by parts, and recollecting that $\delta y = 0$ at both ends, we obtain

$$\delta V = T_1 \int_0^l \frac{dy}{dx}\frac{d\delta y}{dx}\, dx = \tfrac{1}{2}T_1 \int_0^l \delta\left(\frac{dy}{dx}\right)^2 dx,$$

whence

$$V = \tfrac{1}{2}T_1 \int_0^l \left(\frac{dy}{dx}\right)^2 dx \qquad \ldots\ldots\ldots(20).$$

Substituting the value of y from (18) in (20), we obtain

$$V = \tfrac{1}{2}T_1 \int_0^l \{\Sigma_1^\infty \phi_s (s\pi/l) \cos s\pi x/l\}^2 dx$$

$$= \frac{T_1 \pi^2}{4l} \Sigma_1^\infty s^2 \phi_s^2 \qquad \ldots\ldots\ldots(21).$$

Longitudinal Vibrations of Strings.

126. We shall now obtain the equation of motion for the longitudinal vibrations of a string.

Let P and Q be two points whose abscissæ are x, $x + \delta x$; and let these points be displaced to P', Q'. If $x + \xi$ be the abscissa of P', the abscissa of Q' will be $x + \xi + (1 + d\xi/dx)\,\delta x$.

If T be the tension at any point,
$$T = E\frac{P'Q' - PQ}{PQ}.$$
where E is Hooke's modulus of elasticity, whence
$$T = E\frac{d\xi}{dx}.$$
The equation of motion is
$$\rho\delta x \frac{d^2\xi}{dt^2} = \frac{dT}{dx}\delta x + X\rho\delta x$$
and therefore becomes
$$\frac{d^2\xi}{dt^2} = a^2 \frac{d^2\xi}{dx^2} + X,$$
where $a^2 = E/\rho$, and X is the impressed force per unit of mass.

This equation is of the same form as the equation for transverse vibrations, and can be solved in a similar manner.

The conditions to be satisfied at a fixed end are, that the displacement and velocity must be zero throughout the motion; and therefore at a fixed end
$$\xi = 0, \quad \dot\xi = 0$$
for all values of t.

The condition to be satisfied at a free end is that $T = 0$; and therefore at a free end
$$\frac{d\xi}{dx} = 0,$$
for all values of t.

Transverse Vibrations of Membranes.

127. The theory of the vibrations of membranes, is a particular case of the theory of the vibrations of thin elastic plates and shells. In general the stresses across any section of a thin plate or shell consist of [1] (i) a tension T, (ii) a tangential shearing stress M, (iii) a normal shearing stress N, (iv) a flexural couple G, (v) a torsional couple H. If however the membrane is very thin and perfectly flexible, the stresses reduce to a tension T, which in the dynamical problem of small transverse vibrations, may be taken to be equal in all directions, and constant all over the membrane.

We shall now obtain the equation of motion of a plane membrane.

[1] Proc. Lond. Math. Soc. Vol. xxi. p. 33.

VIBRATIONS OF MEMBRANES. 141

Let w be the transverse displacement of any point, the coordinates of whose undisplaced position are $(x, y, 0)$; also let ρ be the superficial density, i.e. the mass of a unit of area of the membrane. If δs, $\delta s'$ be the sides of any small element of the membrane, we may write δx, δy for these quantities; whence the equation of motion is

$$\rho \delta x \delta y \ddot{w} = \frac{d}{dx}\left(T \delta y \frac{dw}{dx}\right) \delta x + \frac{d}{dy}\left(T \delta x \frac{dw}{dy}\right) \delta x,$$

which becomes

$$\frac{d^2 w}{dt^2} = c^2 \left(\frac{d^2 w}{dx^2} + \frac{d^2 w}{dy^2}\right) \quad \ldots\ldots\ldots\ldots\ldots\ldots (22),$$

if $c^2 = T/\rho$.

128. If the boundary of the membrane consists of a rectangle, whose sides are the axes and the lines $x = a$, $y = b$, we may assume as a particular solution of (22),

$$w = A \sin m\pi x/a \sin n\pi y/b \cos pt \ldots\ldots\ldots\ldots (23),$$

where $\quad p^2 = c^2 \pi^2 (m^2/a^2 + n^2/b^2) \ldots\ldots\ldots\ldots\ldots\ldots (24),$

m and n being any integers; for this expression satisfies (22) and also makes w vanish at the boundaries. Equation (24) determines the frequency of the different notes; and from (23) we see that the nodal lines (i.e. the lines of no motion) consist of a system of $n - 1$ lines parallel to x, whose distances apart are b/n, together with $m - 1$ lines parallel to y, whose distances apart are equal to a/m.

If the membrane be square, $a = b$, and (23) and (24) become

$$w = A \sin m\pi x/a \sin n\pi y/a \cos pt,$$

$$p = c\pi (m^2 + n^2)^{\frac{1}{2}}/a.$$

The gravest note is obtained by putting $m = n = 1$, and corresponding to this note there are no nodes.

In the next place we shall determine the nodal lines corresponding to vibrations whose frequency is $\tfrac{1}{2}c\sqrt{5}/a$.

Here $\quad\quad\quad \sqrt{5} = \sqrt{(m^2 + n^2)},$

which requires that $m = 2$, $n = 1$ or $m = 1$, $n = 2$; and therefore the complete vibration corresponding to this period is,

$$w = (C \sin 2\pi x/a \sin \pi y/a + D \sin \pi x/a \sin 2\pi y/a) \cos pt.$$

In this expression C and D depend solely upon the initial circumstances of the motion, and may have any values whatever consistent with the boundary conditions. If however we suppose that the initial conditions are such, that the ratio C/D has an assigned value, we may obtain a variety of special cases.

(i) Let $D = 0$. The nodal system now consists of the line $x = \frac{1}{2}a$, which bisects the membrane.

(ii) Let $C = 0$, and we have a nodal line $y = \frac{1}{2}a$, similarly bisecting the membrane.

(iii) Let $C = D$; then the value of w may be written

$$w = 4C \sin \pi x/a \sin \pi y/a \cos \tfrac{1}{2}\pi (x+y)/a \cos \tfrac{1}{2}\pi (x-y)/a \cos pt.$$

This expression vanishes when,

$$x = a, \quad y = a, \quad x+y = a, \quad x-y = a.$$

The first and second equations correspond to the edges; the fourth must be rejected, because it does not represent a line drawn on the membrane; and the third represents one of the diagonals of the square.

Since a nodal line may be supposed to be rigidly fixed without interfering with the motion, the preceding solution determines the frequency of the gravest note of a right-angled isosceles triangle.

(iv) Let $C = -D$, and we shall find that the nodal line is $y = x$, which represents the other diagonal of the square.

For further examples in this branch of the subject, the reader is referred to Chapter IX. of Lord Rayleigh's treatise.

129. The motion of a circular membrane, which is the best representative of a drum, cannot be solved by elementary methods. The simplest case of all, is when the vibrations are symmetrical with respect to the centre, so that (22) becomes

$$\frac{d^2w}{dt^2} = c^2 \left(\frac{d^2w}{dr^2} + \frac{1}{r}\frac{dw}{dr} \right);$$

and if we put $w = F(r) e^{ipt}$, the equation for F is

$$\frac{d^2F}{dr^2} + \frac{1}{r}\frac{dF}{dr} + p^2 F = 0.$$

This equation cannot be integrated in finite terms. The two solutions are usually called Bessel's functions, from the name of their discoverer; and the investigation of their properties constitutes an important branch of analysis. Algebraic solutions may however be invented, by supposing that the density, and therefore c, is a function of r.

EXAMPLES.

1. A string of length $l+l'$, is stretched with tension T between two fixed points. The linear densities of the lengths l, l' are m, m' respectively; prove that the periods τ of transverse vibrations are given by

$$m'^{\frac{1}{2}} \tan (2\pi l m^{\frac{1}{2}}/\tau T^{\frac{1}{2}}) = m^{\frac{1}{2}} \tan (2\pi l' m'^{\frac{1}{2}}/\tau T^{\frac{1}{2}}).$$

2. Investigate the motion of a string of length l, which is initially at rest in a straight line, each extremity of which is subject to the same obligatory motion $y = k \sin mat$. Show that if a sufficient period be allowed to elapse for the natural vibrations to subside, the position of the nodes will be given by the equation

$$2mx = ml + (2i+1)\pi,$$

where i is any integer.

3. A uniform string in the form of a circle of radius a, rests on a smooth plane under a central repulsion, whose value at distance r is ga^n/r^n. Show that if the string be slightly displaced, so that it is initially at rest and in the form of the curve

$$r = a + \Sigma_1^\infty a_m \cos m\theta,$$

its form at any subsequent time t, will be determined by the equation

$$r = a + \Sigma_1^\infty a_m \cos m\theta \cos m \left\{ \frac{g}{a} \left(\frac{m^2+n-2}{m^2+1} \right) \right\}^{\frac{1}{2}} t.$$

Discuss this result (i) when $m=1$, $n=1$, and (ii) when $n=3$.

4. Three strings OA, OB, OC of the same material but of different lengths, are united at O, and are kept tight by being fastened to fixed points A, B, C, the angles BOC, COA, AOB being denoted by α, β, γ. Show that the times of vibration of the different notes sounded when O is free, are determined by the equation for T, viz.

$$(\sin \alpha)^{\frac{1}{2}} \cot \pi T_1/T + (\sin \beta)^{\frac{1}{2}} \cot \pi T_2/T + (\sin \gamma)^{\frac{1}{2}} \cot \pi T_3/T = 0,$$

where T_1, T_2, T_3 are the times of the gravest notes of OA, OB, OC, when O is fixed.

5. If a stretched string of length l be fastened to two equal masses M, controlled by springs of strength μ allowing transversal vibration, and be plucked at its middle point, prove that the frequency n of vibration will be given by

$$\rho a \tan n\pi l/a = \mu/2n\pi - 2n\pi M,$$

where ρ is the line density, and ρa^2 the tension of the string.

6. A heterogeneous membrane in the shape of a circular annulus, whose edges are fixed and inner and outer radii are b and c, and whose density is μ/r^2, where r is the distance from the centre, is stretched with a tension T, and is performing small symmetrical normal vibrations. Show that a possible motion is given by

$$w = \{A \sin(p \log r/b) + B \sin(p \log c/r)\} \sin(apt + \alpha),$$

where $n\pi = p \log c/b$, n is an integer, and $a^2 = T/\mu$.

7. The fixed boundary of a membrane is square, and the centre of the membrane is displaced perpendicularly through a small space k, the membrane being made to take the form of two portions of intersecting circular cylinders. Prove that the origin being at the centre of the square, the vibrations are given by the equation

$$w = \Sigma A_{nr} \cos \gamma t \sin n\pi (x + a)/2a \sin r\pi (y + a)/2a,$$

where $\qquad 4a^2\gamma^2 = c^2\pi^2(n^2 + r^2).$

Prove that in this case n and r are odd integers, and that

$$A_{nr} = \frac{128k}{\pi^4 (n^2 - r^2)^2} \left(\frac{n^2 + r^2}{nr} - 2 \sin \tfrac{1}{2} n\pi \sin \tfrac{1}{2} r\pi \right),$$

$$A_{nn} = \frac{8k}{n^2\pi^2} \left(1 + \frac{4}{n^2\pi^2} \right).$$

CHAPTER VIII.

FLEXION OF BARS.

130. WE shall now investigate the theory of the equilibrium and the flexural vibrations of a thin rod or bar, and shall confine our attention to two-dimensional motion.

When the bar is not subjected to torsion, the stresses across any section, which are due to the action of contiguous portions of the bar, are completely specified by the following three quantities:—(i) a tension T perpendicular to the section, (ii) a normal shearing stress N, (iii) a flexural couple G. In the figure let PQ be a small element δs, ρ' the radius and O the centre of curvature at P after deformation, σ the density and ω the area of the cross section at P. Also let X, Y, L be the tangential and normal components of the impressed forces and the couple at P, per unit of mass, measured in the directions of T, N, G.

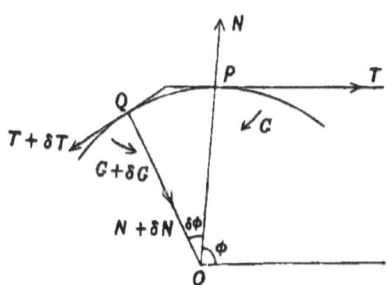

The equations of equilibrium of the bar, are obtained by resolving all the forces along the tangent and normal at P, and taking moments about this point; whence

$$T - (T + \delta T) \cos \delta\phi + (N + \delta N) \sin \delta\phi + \sigma\omega X \delta s = 0,$$
$$N - (T + \delta T) \sin \delta\phi - (N + \delta N) \cos \delta\phi + \sigma\omega Y \delta s = 0,$$
$$G - G - \delta G - (N + \delta N) \rho' \sin \delta\phi + \sigma\omega L \delta s = 0,$$

whence
$$\left. \begin{aligned} \frac{dT}{ds} - \frac{N}{\rho'} &= \sigma\omega X \\ \frac{dN}{ds} + \frac{T}{\rho'} &= \sigma\omega Y \\ \frac{dG}{ds} + N &= \sigma\omega L \end{aligned} \right\} \quad \ldots\ldots\ldots\ldots\ldots\ldots(1).$$

131. We must now find an expression for the flexural couple G.

The curve which passes through the centre of inertia of each cross section, is called the *axis* of the bar. When a bar is bent in such a manner that its curvature is increased, the filaments into which the bar may be conceived to be divided, which lie on the outer side of the axis, will usually be extended; and those which lie on the inner side will usually be contracted, whilst the axis itself undergoes no extension nor contraction. Cases of course may occur, in which the axis undergoes extension or contraction, and when this is the case, the difficulties of the problem are greatly increased, and cannot be satisfactorily discussed without a knowledge of the Theory of Elasticity. We shall therefore confine our attention to the case in which the extension or contraction of the axis is so small (if it exists), that it may be neglected.

In the figure, let AB be the axis of the bar, PQ any filament, whose distance from AB is h, O the centre of curvature at B; also let these points after deformation be denoted by accented letters.

It has been usual to assume, *that the tension T' at P', due to the action of contiguous portions of the bar, is proportional to the extension of the element PQ*; in which case we should have[1]

$$T' = q \frac{P'Q' - PQ}{PQ},$$

---

[1] It would be impossible fully to discuss this assumption in an elementary treatise, but the question may be put in a clearer light by means of the Theory of Elasticity.

where q is a constant called Young's modulus, which depends upon the physical constitution of the rod. Now if ρ, ρ' be the radii of curvature of AB before and after deformation,

$$\frac{PQ}{AB} = \frac{\rho+h}{\rho}, \qquad \frac{P'Q'}{A'B'} = \frac{\rho'+h}{\rho'}.$$

Since we assume that the axis undergoes no extension, $AB = A'B'$; whence

$$T' = q\left(\frac{1+h/\rho'}{1+h/\rho} - 1\right)$$

$$= qh\left(\frac{1}{\rho'} - \frac{1}{\rho}\right),$$

neglecting h^2 etc. Whence

$$G = \iint qh^2\left(\frac{1}{\rho'} - \frac{1}{\rho}\right) dS$$

$$= q\kappa^2\omega\left(\frac{1}{\rho'} - \frac{1}{\rho}\right) \quad\ldots\ldots\ldots\ldots\ldots\ldots(2),$$

where $\kappa^2\omega$ is the moment of inertia of the cross section of the bar.

The quantity of $q\kappa^2\omega$ is sometimes called the *flexural rigidity*, or the *coefficient of flexion*. We shall denote it by E.

Let T'', T''' be the two component tractions perpendicular to the line PQ; σ_1, σ_2, σ_3 the three extensions in the directions T', T'', T'''. Then adopting Thomson and Tait's notation for elastic constants we have

$$T' = (m+n)\sigma_1 + (m-n)(\sigma_2+\sigma_3),$$
$$T'' = (m+n)\sigma_2 + (m-n)(\sigma_3+\sigma_1),$$
$$T''' = (m+n)\sigma_3 + (m-n)(\sigma_1+\sigma_2).$$

Adding the last two, we obtain

$$T'' + T''' = 2m(\sigma_2+\sigma_3) + 2(m-n)\sigma_1,$$

whence
$$T' = \frac{n(3m-n)}{m}\sigma_1 + \frac{m-n}{2m}(T''+T''').$$

If the bar is thin, *and is not subjected to external pressure*, it is probable that the second term in the expression for T' is so small in comparison with the first, that it may be neglected, in which case we should have

$$T' = nm^{-1}(3m-n)\sigma_1.$$

The coefficient $n(3m-n)/m$ is called *Young's modulus*, and is the quantity denoted by q in the text. The assumption therefore supposes that the quantity $(m-n)(T''+T''')/2m$ may be neglected.

The assumption that T'', T''' are *rigorously zero*, which appears to have been made by many writers, is I think unquestionably erroneous except in very special cases; but the retention of these stresses under the above circumstances, would probably lead to terms of a higher order as regards the thickness than $\kappa^2\omega$, and which may be neglected.

The flexural couple G, is therefore proportional to the change of curvature. With regard to the assumption in italics, it is doubtful whether it ought to be made when the surface of the bar is under the action of an external pressure; and we shall therefore suppose that no forces of this description act upon the bar.

In statical problems, the couple L will usually be zero, whilst the forces X, Y will be given; equations (1) together with (2) are therefore sufficient to determine the form of the bar.

132. The conditions to be satisfied at the ends of the bar are the following.

If the ends are subjected to constraining forces and couples, the values of the two stresses T and N at the ends, must be respectively equal to the components along the tangent and normal of the constraining forces; and the couple G, must be equal to the constraining couple.

At a free end, T, N and G must vanish.

133. As an example of these formulæ, we shall consider the Elastica of James Bernoulli, which is the curve assumed by a naturally straight thin bar, whose ends are fastened together by a string of any given length.

Since there are no impressed forces, $X = Y = L = 0$; also $\rho' = ds/d\phi$, whence the first two of (1) become

$$\frac{dT}{d\phi} - N = 0, \quad \frac{dN}{d\phi} + T = 0,$$

and therefore

$$\frac{d^2 T}{d\phi^2} + T = 0;$$

the integral of which is

$$T = A \cos \phi + B \sin \phi,$$

whence $\quad N = -A \sin \phi + B \cos \phi.$

Let t be the tension of the string, and α and $\pi - \alpha$ the values of ϕ at the two extremities, then

$$-t \cos \alpha = N = -A \sin \alpha + B \cos \alpha,$$
$$-t \sin \alpha = T = A \cos \alpha + B \sin \alpha,$$

therefore $\quad A = 0, \quad B = -t.$

THE ELASTICA. 149

Writing $E = q\kappa^2\omega$, and remembering that $\rho = \infty$, since the natural form of the bar is straight, we have $G = E/\rho'$, whence the third of (1) becomes

$$\frac{E}{\rho'}\frac{d}{d\phi}\left(\frac{1}{\rho'}\right) - t\cos\phi = 0 \quad \ldots\ldots \ldots\ldots\ldots\ldots(3).$$

Integrating, we obtain

$$\frac{2E}{\rho'^2} - t\sin\phi = C.$$

Since $G = 0$ when $\phi = \alpha$, $C = -t\sin\alpha$,

whence if $E/t = a^2$
$$\frac{ds}{d\phi} = \frac{a\sqrt{2}}{(\sin\phi - \sin\alpha)^{\frac{1}{2}}} \quad\ldots\ldots\ldots\ldots\ldots\ldots(4),$$

which determines the intrinsic equation to the curve.

We may also integrate (3) in a different manner, for if the string be the axis of x, and its middle point be the origin,

$$\cos\phi = dy/ds,$$

and therefore (3) may be written

$$a^2\frac{d}{ds}\left(\frac{1}{\rho'}\right) = t\frac{dy}{ds},$$

whence $\quad\quad\quad \rho'y = a^2; \quad\ldots\ldots\ldots\ldots\ldots\ldots\ldots(5),$

no constant being required because $\rho'^{-1} = 0$ when $y = 0$.

Now $\quad\quad\quad \rho' = \frac{ds}{dy}\frac{dy}{d\phi} = \frac{dy}{d\phi}\sec\phi,$

whence integrating (5) again

$$y^2 = 2a^2(\sin\phi - \sin\alpha) \quad\ldots\ldots\ldots\ldots\ldots\ldots(6).$$

The forms of the various curves which the wire is capable of assuming, are shown in Thomson and Tait's *Natural Philosophy*, Part II. p. 148.

If α lies between 0 and π, the maxima values of y are obtained by putting $\phi = \frac{1}{2}\pi$, and are therefore equal to $\pm 2a\sin(\frac{1}{4}\pi - \frac{1}{2}\alpha)$. The form of the curve is shown in the figures 1, 2 or 3 of that work; and if the curve be bent upon itself and the slight torsion be neglected, the forms are shown in figures 4 and 5. In all these cases except the first, in which the bar is bent into the shape of a bow, the maximum value of y is numerically equal to its minimum value. If, however, α lies between π and 2π, we may put it equal to $\pi + \beta$, in which case (6) becomes

$$y^2 = 2a^2(\sin\phi + \sin\beta).$$

In this case the maximum value of y occurs when $\phi = \tfrac{1}{2}\pi$, and is equal to $2a \cos(\tfrac{1}{4}\pi - \tfrac{1}{2}\beta)$, and the minimum when
$$\phi = 0, \text{ or } y = a(2\sin\beta)^{\frac{1}{2}};$$
the form of the curve is shown in fig. 7.

The constants a and α are capable of being determined when the lengths of the bar and string are given; and the equation of the curve in Cartesian coordinates can also be obtained, but to do this a knowledge of elliptic functions is required[1]. If, however, $\alpha = \tfrac{3}{2}\pi$, the integral in an algebraic form can be obtained; for since $\tan\phi = -dx/dy$, (6) becomes

$$-x = \int \frac{(y^2 - 2a^2)dy}{y(4a^2 - y^2)^{\frac{1}{2}}}$$
$$= -(4a^2 - y^2)^{\frac{1}{2}} + a \log\{2a/y + (4a^2/y^2 - 1)^{\frac{1}{2}}\} + C.$$

Since $y = a\sqrt{2}$ when $x = 0$, $C = a\sqrt{2} - a\log(\sqrt{2}+1)$,

whence $\quad x = (4a^2 - y^2)^{\frac{1}{2}} - a\sqrt{2} + a\log\dfrac{2a + (4a^2 - y^2)^{\frac{1}{2}}}{y(\sqrt{2}+1)}.$

It will be noticed that this curve is the same as that described by an elliptic cylinder, in the limiting case between oscillation and rotation. See page 71.

134. Equation (4) enables us to prove a theorem discovered by Kirchhoff, and which is known as *Kirchhoff's kinetic analogue*. The theorem is, *that if a point move along the elastica with uniform velocity, the angular velocity of the tangent at that point, is the same as that of a pendulum under the action of gravity.*

If V be the velocity of the moving point, (4) may be written
$$\frac{d\phi}{dt} = \frac{V}{a\sqrt{2}}(\sin\phi - \sin\alpha)^{\frac{1}{2}}.$$

If we put $\quad \chi = \tfrac{1}{2}\pi + \phi, \ \alpha = \tfrac{1}{2}\pi + \beta,$

this becomes $\quad \dfrac{d\chi}{dt} = \dfrac{V}{a\sqrt{2}}(\cos\chi - \cos\beta)^{\frac{1}{2}},$

which is the equation of motion of a common pendulum, whose length is equal $4ga^2/V^2$.

Further information relating to this subject will be found in the following papers[2].

---

[1] Greenhill, *Mess. Math.* vol. VIII. p. 82.

[2] Greenhill, "On the greatest height consistent with stability," *Proc. Camb. Phil. Soc.* vol. IV. p. 65.

Ibid. "On the strength of shafting when exposed both to torsion and end thrust," *Proc. Inst. Mechan. Engineers*, Ap. 1883, p. 182.

Lateral Vibrations.

135. The preceding example illustrates the use of these equations in statical problems; we must now proceed to consider the dynamical theory of the small vibrations of bars.

In order to obtain the equations of motion, we must write $X - \ddot{u}$, $Y - \ddot{v}$ for X and Y in the first two of (1), where u, v, are the tangential and normal displacements; and in the third equation, we must write $L + \kappa^2 \ddot{\phi}$ for L. The equations of motion are thus

$$\left. \begin{aligned} \frac{dT}{ds} - \frac{N}{\rho'} &= \sigma\omega(X - \ddot{u}) \\ \frac{dN}{ds} + \frac{T}{\rho'} &= \sigma\omega(Y - \ddot{v}) \\ \frac{dG}{ds} + N &= \sigma\omega(L + \kappa^2\ddot{\phi}) \end{aligned} \right\} \dots\dots\dots\dots(7).$$

136. We shall now obtain the equation for determining the lateral vibrations of a bar, whose natural form is straight, when under the action of no forces.

In this case $\quad \dfrac{1}{\rho} = 0, \ \dfrac{1}{\rho'} = -\dfrac{d^2v}{dx^2}$.

Since the curvature of the bar is small, ρ'^{-1} is a small quantity; hence *if there is no permanent tension*, the quotient T/ρ' is of the second order of small quantities, and may therefore be neglected; we may also write $-dx$ for ds (ds being measured in the figure in the opposite direction to dx), and the last two of (7) become

$$\frac{dN}{dx} = \sigma\omega\ddot{v} \dots\dots\dots\dots(8),$$

$$q\kappa^2\omega \frac{d^3v}{dx^3} + N = \sigma\kappa^2\omega\dot{\phi} \dots\dots\dots\dots(9).$$

Now $\quad \cot\phi = -dv/dx$,

and since ϕ is very nearly equal to $\tfrac{1}{2}\pi$,

$$\ddot{\phi} = d\ddot{v}/dx,$$

and therefore (9) becomes

$$q\kappa^2\omega \frac{d^3v}{dx^3} + N = \sigma\kappa^2\omega \frac{d^3v}{dt^2 dx} \dots\dots\dots\dots(10);$$

whence eliminating N between (8) and (9) and putting $q/\sigma = b^2$, we obtain

$$\frac{d^2v}{dt^2} + \kappa^2 b^2 \frac{d^4v}{dx^4} - \kappa^2 \frac{d^4v}{dx^2 dt^2} = 0 \quad \ldots\ldots\ldots\ldots(11),$$

which is the required equation of motion.

137. The conditions to be satisfied at a free end are, that G and N should vanish there. It therefore follows from (2) and (10), that these conditions are

$$\frac{d^2v}{dx^2} = 0, \quad \frac{d^3v}{dt^2 dx} - b^2 \frac{d^3v}{dx^3} = 0 \quad \ldots\ldots\ldots\ldots(12).$$

These results agree with those given by Lord Rayleigh, *Theory of Sound*, vol. I. § 162; but the method employed in the text is different, and was suggested by a paper by Dr Besant[1].

138. The third term on the left-hand side of (11) is due to the *rotatory inertia* of the bar, i.e. to say, to the angular motion of the cross sections. This term is generally very small, and it may usually be neglected. When this is the case the equation of motion becomes

$$\frac{d^2v}{dt^2} + \kappa^2 b^2 \frac{d^4v}{dx^4} = 0 \quad \ldots\ldots\ldots\ldots\ldots(13),$$

whilst the boundary conditions at a free end are

$$\frac{d^2v}{dx^2} = 0, \quad \frac{d^3v}{dx^3} = 0 \quad \ldots\ldots\ldots\ldots\ldots(14).$$

139. For the complete discussion of these equations, we must refer the reader to Chapter VIII. of Lord Rayleigh's treatise; but one or two special cases may be noticed.

If the bar is so long that it may be treated as infinite, we may neglect the conditions to be satisfied at its extremities. If therefore the vibrations consist of waves of length λ, we may assume as a solution of (13) that v is proportional to $e^{ipt + 2i\pi x/\lambda}$. Substituting in (13) we obtain,

$$p^2 = 16\pi^4 \kappa^2 b^2/\lambda^4,$$

and therefore the frequency is

$$2\pi \kappa b/\lambda^2.$$

[1] On the Equilibrium of a Bent Lamina, *Quart. Journ.* Vol. IV. p. 12.

When there is a permanent tension T_1, it will be found that we must write $q + T_1$ for q in (2); and the term T_1/ρ' in the second of (7), which becomes $-T_1 \omega d^2v/dx^2$, must be retained. We shall thus obtain the results given by Lord Rayleigh § 188.

140. We shall now investigate the lateral vibrations of a bar[1] of length l.

Taking the origin at the middle point of the bar, we may assume
$$v = U \exp(\iota \kappa b m^2 t/l^2),$$
where U is a function of x, and m is a constant whose value has to be determined. Substituting in (13) we obtain
$$\frac{d^4 U}{dx^4} = \frac{m^4 U}{l^4}.$$

To solve this equation, assume $U = \exp(pmx/l)$, and we see that the values of p are the four fourth roots of unity, viz. $1, -1, \iota, -\iota$. The solution may therefore be written
$$U = A \sin mx/l + B \sinh mx/l$$
$$+ C \cos mx/l + D \cosh mx/l \quad \ldots\ldots\ldots\ldots(15).$$

141. We have now three cases to consider.

(i) Let both ends of the bar be free, so that the bar is what is called a free-free bar. The first of (14) requires that
$$-A \sin mx/l + B \sinh mx/l - C \cos mx/l + D \cosh mx/l = 0,$$
when $x = \pm \tfrac{1}{2} l$. This equation of condition may be satisfied in two different ways; we may first suppose that $C = D = 0$; and
$$-A \sin \tfrac{1}{2} m + B \sinh \tfrac{1}{2} m = 0 \quad \ldots\ldots\ldots\ldots(16),$$
or that $A = B = 0$, and
$$-C \cos \tfrac{1}{2} m + D \cosh \tfrac{1}{2} m = 0 \quad \ldots\ldots\ldots\ldots(17).$$

The first solution corresponds to the first line of (15), which is an odd function of x, and may therefore be called odd vibrations; whilst the second solution corresponds to the second line of (15), which is an even function of x, and may be called even vibrations. We thus see that the odd and even vibrations are independent of one another.

Taking the case of the odd vibrations, the second of (14) requires that
$$-A \cos \tfrac{1}{2} m + B \cosh \tfrac{1}{2} m = 0,$$
and therefore by (16)
$$\tanh \tfrac{1}{2} m = \tan \tfrac{1}{2} m \quad \ldots\ldots\ldots\ldots\ldots(18).$$

---

[1] Greenhill, *Mess. Math.* Vol. XVI. p. 115; Lord Rayleigh, *Theory of Sound*, Ch. VIII.

For the even vibrations, the second of (14) gives
$$C \sin \tfrac{1}{2}m + D \sinh \tfrac{1}{2}m = 0,$$
and therefore by (17)
$$\tanh \tfrac{1}{2}m = -\tan \tfrac{1}{2} m \dots \dots \dots \dots \dots (19).$$

Equations (18) and (19) determine the values of m for the odd and even vibrations respectively, and consequently the frequency of the different notes can be found.

(ii) Let both ends of the bar be clamped, so that the bar is clamped-clamped.

In this case the conditions to be satisfied at the ends are, that
$$U = 0, \quad dU/dx = 0 \dots \dots \dots \dots \dots (20),$$
the first of which expresses the condition that the displacement at each end should be zero, and the second that the direction of the axis should be unchanged.

The solution for this case may evidently be obtained by integrating the results for a free-free bar twice with respect to x, and consequently the values of m for the odd and even vibrations are given by (18) and (19).

(iii) Let the bar be clamped-free, i.e. clamped at $x = -\tfrac{1}{2}l$, and free at $x = \tfrac{1}{2}l$.

When $x = -\tfrac{1}{2}l$ equations (20) have to be satisfied; and when $x = \tfrac{1}{2}l$, the conditions are given by (14). Taking the value of U given by (15) and writing out the four equations of condition in full, it will be found that they can be satisfied in two ways, i.e. either $B = C = 0$,
$$-A \sin \tfrac{1}{2}m + D \cosh \tfrac{1}{2}m = 0,$$
$$A \cos \tfrac{1}{2}m - D \sinh \tfrac{1}{2}m = 0,$$
which gives
$$\tanh \tfrac{1}{2}m = \cot \tfrac{1}{2}m \dots \dots \dots \dots \dots (21),$$
or, $A = D = 0$,
$$-B \sinh \tfrac{1}{2}m + C \cos \tfrac{1}{2}m = 0,$$
$$B \cosh \tfrac{1}{2}m + C \sin \tfrac{1}{2}m = 0,$$
which gives
$$\tanh \tfrac{1}{2}m = -\cot \tfrac{1}{2}m \dots \dots \dots \dots \dots (22).$$

Equations (18) and (19) are both included in the equation
$$\cos m \cosh m = 1 \dots \dots \dots \dots \dots (23),$$

and (21) and (22) in the equation

$$\cos m \cosh m = -1 \ldots\ldots\ldots\ldots\ldots(24).$$

For a discussion of the roots of these equations, we must refer to Lord Rayleigh's *Theory of Sound*, Chapter VIII. and to Prof. Greenhill's paper.

Longitudinal Vibrations.

142. The equation of motion for longitudinal vibrations of a straight bar may be obtained immediately from the first of (7). In this case $\rho' = \infty$, $ds = -dx$, and therefore

$$\frac{dT}{dx} = \sigma\omega \frac{d^2u}{dt^2},$$

u being the longitudinal displacement.

But $$T = q\omega \frac{du}{dx},$$

whence putting $q/\sigma = b^2$, we obtain

$$\frac{d^2u}{dx^2} = b^2 \frac{d^2u}{dt^2},$$

which is the same equation which we have obtained for the lateral vibrations of a string. The condition to be satisfied at a fixed end is that $u = 0$; whilst the condition to be satisfied at a free end is that $T = 0$, or $du/dx = 0$.

In the case of a bar of infinite length, which propagates waves of length λ, we must put

$$u = e^{2i\pi x/\lambda + \iota pt},$$

and therefore $$p = 2\pi b/\lambda = \frac{2\pi}{\lambda}\sqrt{\frac{q}{\sigma}}.$$

In the corresponding case of lateral vibrations of the same wave-length,

$$p' = \frac{2\pi\kappa}{\lambda^2}\sqrt{\frac{q}{\sigma}},$$

whence $$p'/p = \kappa/\lambda.$$

Since λ is usually very much greater than κ, which is the radius of gyration of the cross section, we see that the pitch of notes arising from longitudinal vibrations, is usually much higher than that of notes arising from lateral vibrations.

Vibrations of a Circular Bar.

143. We shall not discuss the theory of bars whose natural form is curved, but there is one result which can be obtained without difficulty, viz. the frequency of the vibrations of a bar which forms a complete circle of radius a.

In this case we may in equations (7) put $ds = ad\phi$, and $\rho' = a$, since the difference between ρ'^{-1} and a^{-1} may be neglected when multiplied by T or N. Equations (7) therefore become, measuring u in the opposite direction, so that u and ϕ increase together,

$$\left. \begin{array}{l} \dfrac{dT}{d\phi} - N = \sigma a \omega \ddot{u} \\[4pt] \dfrac{dN}{d\phi} + T = -\sigma a \omega \ddot{v} \\[4pt] \dfrac{dG}{d\phi} + Na = 0 \end{array} \right\} \quad \ldots\ldots\ldots\ldots(25),$$

the rotatory inertia being neglected.

We must now find an expression for the change of curvature due to deformation.

If R, Φ be the coordinates after displacement, of the point on the axis which was initially at (a, ϕ), then

$$R = a + v, \quad \Phi = \phi + u/a.$$

Hence if P be the perpendicular from the centre on to the tangent to the deformed axis, at the point in question, we have by a well-known formula,

$$\frac{1}{\rho'} = \frac{1}{R}\frac{dP}{dR}.$$

Now
$$\frac{1}{P^2} = \frac{1}{R^2}\left\{1 + \frac{1}{R^2}\left(\frac{dR}{d\Phi}\right)^2\right\}$$

$$= \frac{1}{R^2}\left\{1 + \frac{(dv/d\phi)^2}{R^2(1 + du/ad\phi)^2}\right\};$$

also the displacements and their differential coefficients are all small quantities; whence expanding and neglecting cubes of small quantities, the above equation becomes

$$P = a + v - \frac{1}{2a}\left(\frac{dv}{d\phi}\right)^2,$$

whence
$$dP = \left(1 - \frac{1}{a}\frac{d^2v}{d\phi^2}\right)\frac{dv}{d\phi}d\phi.$$

Also $dR = \dfrac{dv}{d\phi} d\phi,$

therefore $\dfrac{1}{\rho'} - \dfrac{1}{a} = -\dfrac{1}{a^2}\left(\dfrac{d^2v}{d\phi^2} + v\right)$ ....(26),

which determines the change of curvature in terms of the normal displacement.

We must next find the condition that the axis undergoes no extension.

The elementary arc ds' of the deformed surface is given by the equation
$$ds'^2 = (dv)^2 + (a+v)^2 (d\phi + du/a)^2,$$
and since this is equal to $a^2 d\phi^2$, we obtain, neglecting squares of small quantities,
$$\dfrac{du}{d\phi} + v = 0 ....(27),$$
which is the condition of inextensibility.

Substituting from (26) and (2) in the last of (25) we obtain
$$\dfrac{q\kappa^2 \omega}{a^3}\left(\dfrac{d^2v}{d\phi^2} + \dfrac{dv}{d\phi}\right) = N.$$

From the first two of (25) we obtain
$$\dfrac{d}{d\phi}\left(\dfrac{d^2N}{d\phi^2} + N\right) = -\sigma a\omega \left(\dfrac{d^2v}{d\phi^2} + \dfrac{du}{d\phi}\right)$$
$$= -\sigma a\omega \left(\dfrac{d^2\ddot{v}}{d\phi^2} - \ddot{v}\right),$$
by (25), whence eliminating N we obtain
$$\dfrac{q\kappa^2}{\sigma a^4}\left(\dfrac{d^2}{d\phi^2} + 1\right)^2 \dfrac{d^2v}{d\phi^2} + \left(\dfrac{d^2}{d\phi^2} - 1\right)\ddot{v} = 0 \ ....(28).$$

To solve this equation, assume that $v \propto \epsilon^{\iota pt + u\phi}$, and we obtain
$$p^2 = \dfrac{q\kappa^2 s^2(s^2-1)^2}{\sigma a^4(s^2+1)} \ ....(29).$$

This result was first obtained by Hoppe[1].

If the bar is a complete circle, v must necessarily be periodic with respect to ϕ, and therefore s must be an integer, unity and zero excluded. We therefore see that there are an infinite number of modes of vibration, whose frequencies are obtained by putting $s = 2, 3, 4...$ in (29).

[1] Crelle, Vol. LXIII.; and see Lord Rayleigh, *Theory of Sound* § 233.

If the bar is not a complete circle, s is not an integer; its values in terms of p are the six roots of (29), but since p is unknown, another equation is necessary. This equation is obtained by considering the boundary conditions to be satisfied at the free ends, and which are that T, N and G should vanish there. These conditions will furnish six additional equations, by means of which the six constants which appear in the solution of (28) can be eliminated, and the resulting determinantal equation combined with (29), will determine the frequency[1].

EXAMPLES.

1. A naturally straight bar AB, of which the end A is fixed, is lying on a smooth horizontal plane, and the other end is pulled with a force F, whose direction is perpendicular to the undisplaced position of the bar. Prove that the projection of any length AP on the undisplaced position AB, is equal to

$$(2F/E)^{\frac{1}{2}} \{\sqrt{(\cos \beta)} - \sqrt{(\cos \beta - \cos \phi)}\},$$

where ϕ is the angle which the normal at P makes with AB, and β is the value of ϕ at the end B.

2. If a uniform horizontal bar, both of whose ends are fixed, be displaced horizontally, so that one half is uniformly extended, and the other half is uniformly compressed, prove that the displacement at time t of any particle whose abscissa is x, is

$$(4nl/\pi^2) \Sigma (2i+1)^{-2} \cos (2i+1) \pi at/2l \cos (2i+1) \pi x/2l,$$

where $2l$ is the length of the bar, and the middle of which is the origin, and nl is the initial displacement of that point.

3. The extremities of a uniform bar of length l, are attached to two fixed points distant l apart by springs of equal strength. Show that if the longitudinal displacement of the bar is represented by $P e^{\iota mat} \sin (mx/l + \alpha)$, the admissible values of m are given by the equation

$$(m^2 q^2 - l^2 \mu^2) \tan m + 2mql\mu = 0,$$

where μ is the strength of either of the springs, and q the ratio of the tension to the extension in the bar.

---

[1] See Lamb, "On the flexure and the vibrations of a curved bar," *Proc. Lond. Math. Soc.* Vol. xix. p. 365.

4. An elastic wire, indefinitely extended in one direction, is firmly held in a clamp at the other end. If a series of simple transverse waves travelling along the wire be reflected at the clamp; show that the reflected waves will have the same amplitude as the incident waves, but that their phase is accelerated by one quarter of a wave length.

5. A heavy wire of uniform section is carried on a series of supports in the same horizontal plane, L_r is the bending moment at the rth point of support, l_r the distance between the $(r-1)$th and the rth support, and m the mass of the wire per unit of length; prove that

$$L_{r-1}l_r + 2L_r(l_r + l_{r+1}) + L_{r+1}l_{r+1} = \tfrac{1}{4}mg(l_r^3 + l_{r+1}^3).$$

6. Prove that if an elastic bar of length l with flat ends, impinges directly with velocity V on a longer bar at rest, of length nl and of the same material and cross section, also with flat ends, the first bar will be reduced to rest by the impact; and the second bar will appear to move with successive advances of the ends with velocity V for intervals of time $2l/a$, and intervals of rest of $2(n-1)l/a$, a denoting the velocity of propagation of longitudinal vibrations.

7. An elastic rod of length l lies on a smooth plane, and is longitudinally compressed between two pegs at a distance l' apart. One peg is suddenly removed; prove that the rod leaves the other peg just as it reaches its natural state, and then proceeds with a velocity equal to $V(l-l')/l$, where V is the velocity of propagation of a longitudinal wave in the rod.

8. A metal rod fits freely in a tube of the same length, but of a different substance, and the extremities of each are united by equal perfectly rigid discs fitted symmetrically at the end. Show that the frequencies of the notes emissible, which have a node at the centre of the system, are given by $x/2\pi l$, where $2l$ is the length of the rod or tube, and n is a root of the equation

$$2Mx = ma \cot x/a + m'a' \cot x/a';$$

where M, m, m' are the masses of a disc, the bar, and the tube, and a, a' are the velocities of propagation of sound along the bar and the tube.

9. Two equal and similar elastic rods AC, BC are hinged at C so as to form a right angle, while their other extremities are clamped. One vibrates transversely and the other longitudinally; prove that the periods are $2l^2/f^2\theta^2$, where θ is given by the equation

$$1 + \cosh\theta \cos\theta$$
$$+ (\sin\theta \cosh\theta - \cos\theta \sinh\theta)(gl/f^2\theta)\cot(\theta^2 f^2/gl) = 0,$$

where l is the length of either rod, and f, g are two constants depending on the material.

10. The natural form of a thin rod when at rest is a circular arc, and the rod makes small oscillations about this form in its own plane. Assuming that the couple due to bending varies as the change of curvature, and that the tension follows Hooke's law, prove that if the arc be a complete circle, the periods $2\pi/p$ are given by the quadratic,

$$p^4 - \{b(n^2+1) + an^2(n^2-1)\}p^2 + abn^2(n^2-1) = 0,$$

where n is any integer, and a, b are two constants which depend upon the moduli of stretching and bending, and on the radius of the circle.

11. If in the last example the arc be not a complete circle, but have both ends free and be inextensible, show that it can be made to vibrate symmetrically about its middle point by suitable initial conditions in a period $2\pi/p$, provided the angle 2θ which the arc subtends at its centre, satisfies the equation

$$q(q^2+1)(q'^2-q''^2)\cot q\theta + q'(q'^2+1)(q''^2-q^2)\cot q'\theta$$
$$+ q''(q''^2+1)(q^2-q'^2)\cot q''\theta = 0,$$

where q^2, q'^2, q''^2 are the roots, real or imaginary, of the cubic

$$ax(x^2-1)^2 = (x+1)p^2.$$

CHAPTER IX.

EQUATIONS OF MOTION OF A PERFECT GAS.

144. WE have already called attention to the fact, that air is the vehicle by means of which sound is transmitted; we must therefore investigate the equations of motion of a gas.

The general equations of fluid motion, which we obtained in Chapter I, are of course applicable to elastic fluids such as air and other gases, as well as to incompressible fluids such as water; but in order to investigate the propagation of sound in gases, these equations require modification.

In all problems relating to vibrations, the velocities upon which the vibrations depend, are usually so small that their squares and products may be neglected; also the variation of the density of the gas is usually a small quantity. If therefore a gas, which is at rest, be disturbed by the passage of sound waves, we may write d/dt for $d/dt + u\,d/dx + v\,d/dy + w\,d/dz$, and also put $\rho = \rho_0(1+s)$, where s, which is called the condensation, is a small quantity. The equations of motion therefore become

$$\left. \begin{aligned} \frac{du}{dt} &= X - \frac{1}{\rho}\frac{dp}{dx} \\ \frac{dv}{dt} &= Y - \frac{1}{\rho}\frac{dp}{dy} \\ \frac{dw}{dt} &= Z - \frac{1}{\rho}\frac{dp}{dz} \end{aligned} \right\} \quad \ldots\ldots\ldots\ldots(1),$$

whilst the equation of continuity, § 6, equation (5), becomes

$$\frac{ds}{dt} + \frac{du}{dx} + \frac{dv}{dy} + \frac{dw}{dz} = 0 \ldots\ldots\ldots\ldots(2).$$

B. II.

We shall also suppose that the bodily forces (if any) which act upon the gas arise from a potential U, and also that the motion is irrotational; (2) therefore becomes

$$\frac{ds}{dt} + \nabla^2\phi = 0 \quad\quad\quad\quad\quad (3).$$

We have already shown that when the motion is irrotational, the pressure is determined by the equation

$$\int\frac{dp}{\rho} + U + \frac{d\phi}{dt} + \tfrac{1}{2}q^2 = C \quad\quad\quad\quad\quad (4).$$

Now q^2 is to be neglected, also if we assume Boyle's law to hold, we shall have

$$p = k\rho = k\rho_0(1+s),$$

and therefore

$$\int\frac{dp}{\rho} = k\int\frac{ds}{1+s} = k\log(1+s) + C',$$
$$= ks + C',$$

neglecting s^2 &c. Whence (4) becomes

$$ks + U + \dot\phi + C' = C.$$

If there were no forces in action and no motion, the first three terms would be zero; whence $C' = C$, and therefore,

$$ks + U + \dot\phi = 0 \quad\quad\quad\quad\quad (5),$$

or if δp denote the small variable part of p, (5) may be written

$$\frac{\delta p}{\rho_0} + U + \dot\phi = 0 \quad\quad\quad\quad\quad (6).$$

Eliminating s between (3) and (5) we obtain

$$\frac{d^2\phi}{dt^2} = k\nabla^2\phi - \frac{dU}{dt} \quad\quad\quad\quad\quad (7).$$

Equation (6) and (7) are the fundamental equations of the small vibrations of a gas.

145. In almost all the applications of these equations, no impressed forces act, and therefore $U = 0$; accordingly (7) becomes

$$\frac{d^2\phi}{dt^2} = k\nabla^2\phi \quad\quad\quad\quad\quad (8).$$

Let us now suppose that plane waves of sound are propagated in a gas of unlimited extent. Let l, m, n be the direction cosines

of the wave front, a the velocity of propagation of the wave. We may assume,
$$\phi = A\epsilon^{\iota\kappa(lx+my+nz-at)}.$$

Substituting in (8) we obtain
$$k = a^2 \dots\dots\dots\dots\dots\dots\dots\dots(9).$$

This equation determines the physical meaning of k, and shows that it is equal to the square of the velocity of propagation. We may therefore write (8) in the form
$$\frac{d^2\phi}{dt^2} = a^2\nabla^2\phi\dots\dots\dots\dots\dots\dots(10).$$

Let ξ, η, ζ be the displacements of an element of fluid, then
$$\frac{d\xi}{dt} = \frac{d\phi}{dx} = A\iota\kappa l\epsilon^{\iota\kappa(lx+my+nz-at)},$$
whence
$$\xi = -(Al/a)\,\epsilon^{\iota\kappa(lx+my+nz-at)},$$
with similar expressions for η and ζ. We thus obtain
$$\xi/l = \eta/m = \zeta/n,$$
which shows that the displacement is *perpendicular* to the front of the wave. This constitutes one of the fundamental distinctions between sound waves and waves of light, for it is well known that in a wave of light the direction of displacement always lies *in* the wave front. It therefore follows that sound waves are incapable of polarization; they are, however, capable of interfering with one another and also of being diffracted, since these phenomena do not depend upon the direction of vibration.

146. Equation (9) enables us to calculate the velocity of sound in a gas, and we shall now show how it may be applied to obtain the velocity of sound in air.

We have
$$a = \sqrt{k} = \sqrt{(p/\rho)},$$
where p is the pressure corresponding to a given density. Now it is found by experiment that at $0°$ C. under a pressure equal to the weight of 1033 grammes per square centimetre, at the place where the experiment is made (i.e. a pressure equal to 1033 g barads[1]), the density of dry air is ·001293 grammes per cubic

---

[1] In the report of the British Association at Bath, 1888, the Committee on Units recommended the introduction of the following additional units, viz. that

(i) The unit of velocity on the c. g. s. system, i.e. the velocity of one centimetre per second, should be called one *kine*.

centimetre. Hence if we employ the C. G. S. system units, and take $g = 981$, we obtain
$$p = 1033\, g = 1033 \times 981, \quad \rho = \cdot 001293,$$
which gives
$$a = 27995;$$
so that the velocity of sound at $0°$ C. is $279 \cdot 95$ metres per second, or $918 \cdot 49$ feet per second.

The first theoretical investigation respecting the velocity of sound in air was made by Newton, but when his result was submitted to experiment, it was found that it was too small by about one-sixth, in as much as the correct result is about 1089 feet per second. This discrepancy between theory and observation was not explained for more than a century, until Laplace pointed out, that the use of Boyle's law involved the assumption, that the temperature remains constant throughout the motion, whereas it is well known that when a gas is suddenly compressed its temperature rises. Now it was supposed by Laplace that in the case of sound waves, the condensation and rarefaction take place so suddenly, that the heat or cold produced have not time to disappear by conduction, and consequently the motion which takes place is much the same as it would be, if the air were confined in a non-conducting vessel. We must therefore ascertain the relation between the pressure and density under these circumstances, and shall accordingly make a short digression on the Thermodynamics of Gases.

Thermodynamics of Gases[1].

147. Let us suppose that a unit mass of gas is contained in a cylinder filled with a moveable piston, and let p, v, E be its pressure volume and intrinsic energy. Also let θ be the temperature measured from the absolute zero of the air thermometer, i.e. $-273°$ C.

(ii) The unit of momentum, i.e. the momentum of one gramme moving with the velocity of one kine, should be called one *bole*.

(iii) The unit of pressure, i.e. the pressure of one dyne per square centimetre, should be called one *barad*.

When employing absolute units, it is most important to recollect, that a gramme represents a unit of *mass* and not a unit of *weight*.

[1] The reader is supposed to have studied some elementary work on Thermodynamics, such as Maxwell's *Heat*.

Let a small quantity dH of heat (expressed in mechanical units) be communicated to the gas. If the gas be allowed to expand, the effect of this heat will be (i) to do an amount of work which is equal to pdv, and (ii) to increase the intrinsic energy by dE. Now the first law of Thermodynamics asserts that— *When work is transformed into heat, or heat into work, the quantity of work is mechanically equivalent to the quantity of heat.* It therefore follows from this law, that

$$dE = dH - pdv \quad\quad\quad\quad\quad (11).$$

By virtue of the laws of Boyle and Charles, the relation

$$pv = h\theta \quad\quad\quad\quad\quad (12)$$

exists between the pressure, volume and temperature of a gas. Any two of the quantities p, v or θ may accordingly be taken as the independent variables. If therefore we take v and θ as independent variables, we may write

$$dH = ldv + K_v d\theta \quad\quad\quad\quad\quad (13).$$

The quantity l is the *latent heat of expansion*, and K_v is the *specific heat at constant volume*, both expressed in mechanical units.

It is important to notice that the right hand-side of (13) is not a perfect differential; for although dH is *in form* the differential of a quantity dH of heat, yet it is not a definite function of the volume and temperature. The amount of heat communicated to a substance may be measured in mechanical or thermal units, but it cannot be regarded as a function of the state of the substance to which it is communicated.

The intrinsic energy on the other hand, is a function of the state of the substance, and therefore dE is the differential of a definite function of any two of the quantities p, v, θ.

Equations (11) and (13) are therefore equivalent to

$$dE = (l - p)\, dv + K_v d\theta \quad\quad\quad\quad\quad (14).$$

This equation is true of all substances; but the experiments of Joule and Sir W. Thomson have shown that, *the intrinsic energy of a unit of mass of a perfect gas is almost entirely dependent upon its temperature, and not upon its volume.* Accordingly E is a function of θ and not of v, and therefore

$$\frac{dE}{dv} = 0, \quad\quad \frac{dE}{d\theta} = F'(\theta).$$

From these equations combined with (14) we obtain

$$l = p, \quad K_v = F(\theta)$$

which shows that the latent heat of expansion is equal to the pressure, and that the specific heat at constant volume is a function of the temperature.

Equation (13) may therefore be written

$$dH = p\,dv + F(\theta)\,d\theta,$$

whence by (12)

$$\frac{dH}{\theta} = \frac{h\,dv}{v} + \frac{F(\theta)}{\theta}\,d\theta \quad\ldots\ldots\ldots\ldots\ldots\ldots(15).$$

The right-hand side of this equation is a perfect differential of a function which we shall denote by ϕ, accordingly (15) may be written

$$dH = \theta\,d\phi \ldots\ldots\ldots\ldots\ldots\ldots\ldots(16).$$

148. This equation is the analytical expression of a very important but somewhat recondite law, known as the second law of Thermodynamics. For a full discussion of the second law, we must refer to treatises on Thermodynamics, but a few remarks on this subject may be useful.

If one substance at a temperature S be placed in contact with another substance at a lower temperature T, heat will flow from the hot substance into the cold substance; and this process will continue until both substances are reduced to the same temperature. It can however be shown by means of a theoretical heat engine devised by Carnot, that it is possible to transfer heat from a cold body to a hot body by means of the expenditure of work; and the second law asserts, that it is impossible to do this *without expenditure of work*. The law was first enunciated by Clausius in the following terms:—

It is impossible for a self-acting machine, unaided by external agency, to convey heat from one body to another at a higher temperature.

Sir W. Thomson states the law in a slightly different form as follows:—

It is impossible by means of inanimate material agency, to derive mechanical effect from any portion of matter, by cooling it below the temperature of the coldest surrounding objects.

By means of the experimental law, that the intrinsic energy of a gas depends upon its temperature and not upon its volume, the second law of Thermodynamics may be dispensed with in dealing with gases; or to put the matter more correctly, the second law can be deduced as a consequence of the experimental law. But in the case of substances which are not in the gaseous state, the first law is not sufficient to enable us to investigate their thermodynamical properties. Moreover, although it is always assumed that the pressure, temperature and volume are connected together by a certain relation, which may be mathematically expressed by an equation of the form $F(p, v, \theta) = 0$; yet the form of the function F is not accurately known, except in the case of perfect gases. It can be shown that for all substances the second law is mathematically expressed by means of equation (16), and it thus leads to a certain function ϕ, which is capable of being theoretically expressed as a function of any two of the quantities p, v, θ, and which specifies the properties of the substance when it is not allowed to gain or lose heat.

The function ϕ was called the Thermodynamic Function by Rankine; but it is now always known as the *Entropy*.

149. Returning to § 147, let us take p and θ to be the independent variables; equation (13) may then be written

$$dH = Rdp + K_p d\theta,$$

where K_p is the specific heat at constant pressure. Substituting in (11) and eliminating dp by (12) we obtain

$$dE = (Rp/\theta + K_p) d\theta - p(1 + R/v) dv.$$

Since the right-hand side of this equation must be identical with the right-hand side of (14), we must have

$$R = -v, \quad Rp/\theta + K_p = K_v;$$

whence
$$K_p - K_v = h \quad \ldots\ldots\ldots\ldots\ldots\ldots\ldots(17).$$

Equation (17) shows that the *difference* between the two specific heats is constant; also since the specific heat at constant volume has been shown to be a function of the temperature, it follows that the specific heat at constant pressure must also be a function of the temperature.

150. The value of the specific heat of air at constant pressure has been determined by Regnault, and he finds that it is very nearly independent of the temperature, and is equal to 183·6 foot-

pounds[1] per degree Fahrenheit. It therefore follows from (17) that K_v is also very nearly independent of the temperature.

It also follows from Regnault's experiments, that the value of h for air is 53·21 foot-pounds per degree Fahrenheit; we thus obtain
$$K_v = K_p - h,$$
$$= 183·6 - 53·21 = 130·4.$$

The quantity with which we are most concerned in Acoustics, is the ratio of the specific heat at constant pressure, to the specific heat at constant volume, which is usually denoted by γ. We accordingly find
$$\gamma = K_p/K_v = 1·408.$$

The specific heats of all perfect gases are so very nearly independent of the temperature, that they may be treated as constant. The value of the ratio γ, is also approximately the same for all gases.

151. We have already proved the equation $dH = \theta d\phi$. The quantity ϕ is called by Clausius the *entropy of the gas*, and is a quantity which specifies in an analytical form, the properties of a gas which expands or contracts without loss or gain of heat; for when this is the case $dH = 0$, and therefore $\phi = $ const. If therefore we suppose that ϕ is expressed as a function of p and v, the curve $\phi = $ a const. on the indicator diagram, will be a curve which represents the state of the gas under these circumstances. Such curves are called *adiabatic lines*, or *isentropic lines*.

In order to find the form of these curves, we must find an expression for the entropy. Remembering that $l = p$, we obtain from (13) and (16)
$$\theta d\phi = p dv + K_v d\theta \quad \dots\dots\dots\dots\dots\dots(18),$$
$$= \frac{h\theta}{v} dv + K_v d\theta,$$
whence
$$\phi = h \log v + K_v \log \theta + \text{const.} \dots\dots\dots\dots(19).$$

By (12) and (17), this may be expressed in the form
$$\phi = (K_p - K_v) \log v + K_v \log pv/h + \text{const.},$$
whence
$$pv^\gamma = A \epsilon^{\phi/K_v} \quad \dots\dots\dots\dots\dots\dots\dots(20),$$
where A is a constant.

This is the equation of the adiabatic lines of a perfect gas.

---

[1] This calculation is taken from Chapter XI. of Maxwell's *Heat*, in which British units are employed.

If ρ be the density of the gas, $v \propto \rho^{-1}$; whence by (20) the relation between the pressure and density of a gas, which expands without loss or gain of heat, is

$$p = k'\rho^\gamma \quad\ldots\ldots\ldots\ldots\ldots\ldots\ldots(21),$$

where k' is a constant.

The equation of the isothermal lines may be written

$$pv = (K_p - K_v)\theta \quad\ldots\ldots\ldots\ldots\ldots\ldots(22).$$

152. The mechanical properties of perfect gases are specified by two quantities, viz. their *densities* and their *elasticities*. The density, as is well known, is defined to be the mass of a unit of volume; but in order to understand what is meant by the elasticity of a gas, some further definitions will be necessary.

The elasticity of a gas under any given conditions, is the ratio of any small increase of pressure, to the voluminal compression thereby produced.

The voluminal compression, is the ratio of the diminution of volume to the original volume.

Hence if v the original volume, be reduced by the application of pressure δp, to $v + \delta v$ (δv being of course negative), the elasticity E is equal to

$$E = -v\frac{dp}{dv} = \rho\frac{dp}{d\rho} \quad\ldots\ldots\ldots\ldots\ldots\ldots(23).$$

The quantity E is called the compressibility by Lord Rayleigh (Chapter XV.), and is denoted by him by m.

The value of $dp/d\rho$, and therefore E, depends upon the thermal conditions under which the compression takes place. The two most important cases are, (i) when the temperature remains constant, (ii) when there is no loss or gain of heat. We shall, following Maxwell, denote the elasticity under these two conditions by E_θ and E_ϕ.

In the first case $p = k\rho$, whence $dp/d\rho = k$; accordingly

$$E_\theta = k\rho = p \quad\ldots\ldots\ldots\ldots\ldots\ldots(24).$$

In the second case $p = k'\rho^\gamma$, whence $dp/d\rho = k'\gamma\rho^{\gamma-1}$; accordingly

$$E_\phi = k'\gamma\rho^\gamma = \gamma p \quad\ldots\ldots\ldots\ldots\ldots\ldots(25).$$

From (24) and (25) we obtain,

$$\frac{E_\phi}{E_\theta} = \gamma = \frac{K_p}{K_v} \quad\ldots\ldots\ldots\ldots\ldots\ldots(26).$$

Velocity of Sound in Air.

153. Having made this digression upon the thermodynamics of gases, we are prepared to investigate the velocity of sound in a gas.

From (21) we obtain

$$\int \frac{dp}{\rho} = \frac{k'\gamma}{\gamma - 1} \rho^{\gamma-1}$$

$$= \frac{k'\gamma}{\gamma - 1} \rho_0^{\gamma-1} + k'\gamma \rho_0^{\gamma-1} s,$$

since $\rho = \rho_0 (1 + s)$. Whence (4) becomes

$$k'\gamma \rho_0^{\gamma-1} s + U + \dot{\phi} = 0.$$

Eliminating s from (3) we obtain

$$\frac{d^2\phi}{dt^2} = k'\gamma \rho_0^{\gamma-1} \nabla^2 \phi - \frac{dU}{dt},$$

and therefore the velocity of sound is equal to $(k'\gamma \rho_0^{\gamma-1})^{\frac{1}{2}}$.

Now $\quad k = p_0/\rho_0,$

and $\quad k' = p_0/\rho_0^\gamma,$

whence $\quad k'\rho_0^{\gamma-1} = k.$

The velocity of sound is therefore equal to $(k\gamma)^{\frac{1}{2}}$ and is therefore augmented in the ratio $\sqrt{\gamma} : 1$. In the case of air, the value of $k^{\frac{1}{2}}$ in feet per second has already been shown to be equal to 918·49, and therefore

$$(k\gamma)^{\frac{1}{2}} = 1083·82,$$

which nearly agrees with the value 1089 feet per second given above.

Intensity of Sound[1].

154. We have stated in § 111 that the intensity of sound is measured by the rate at which energy is transmitted across unit area of the wave front. We shall therefore find an expression for this quantity.

Let the velocity potential of a plane wave be

$$\phi = A \cos \frac{2\pi}{\lambda} (x - Vt),$$

---

[1] Lord Rayleigh, *Theory of Sound*, § 245.

then
$$\frac{d\phi}{dx} = -\frac{2\pi A}{\lambda}\sin\frac{2\pi}{\lambda}(x - Vt).$$

If p_0, $p_0 + \delta p$ be the pressures when the air is at rest and in motion respectively, the rate dW/dt at which work is transmitted is
$$\frac{dW}{dt} = (p_0 + \delta p)\frac{d\phi}{dx};$$

and since
$$\delta p = -\rho\dot\phi = -\rho_0 V\frac{2\pi A}{\lambda}\sin\frac{2\pi}{\lambda}(x - Vt),$$

we obtain
$$\frac{dW}{dt} = -\left\{p_0 - \rho_0 V\frac{2\pi A}{\lambda}\sin\frac{2\pi}{\lambda}(x - Vt)\right\}\frac{2\pi}{\lambda}\sin\frac{2\pi}{\lambda}(x - Vt),$$
$$= \frac{2\pi^2 A^2 \rho_0}{V\tau^2} + \text{periodic terms,}$$

since $V\tau = \lambda$.

We therefore see that the rate at which energy is transmitted, consists of two terms: viz. a constant term, which shows that a definite quantity of energy flows across the wave front per unit of time; and a periodic term, which fluctuates in value and contributes nothing to the final effect. The first term measures the intensity of sound, and shows that it varies directly as the square of the amplitude, and inversely as the product of the velocity of propagation in the medium and the square of the period.

CHAPTER X.

PLANE AND SPHERICAL WAVES.

155. We shall devote the present chapter to the consideration of certain special problems relating to plane and spherical waves of sound.

The theory of the vibrations of strings, which was discussed in Chapter VII., explains the production of notes by means of stringed instruments; but in order to understand how notes are produced by means of wind instruments, it will be necessary to investigate the motion of air in a closed or partially closed vessel. The simplest problem of this kind is the motion of plane waves of sound in a cylindrical pipe, which we shall proceed to consider.

Motion in a Cylindrical Pipe.

156. Let l be the length of a cylindrical pipe, whose cross section is any plane curve, and let the fronts of the waves be perpendicular to the sides of the cylinder.

We shall suppose for simplicity, that the motion is in one dimension, whence measuring x from one end of the pipe, the equation of motion is

$$\frac{d^2\phi}{dt^2} = a^2 \frac{d^2\phi}{dx^2} \quad\ldots\ldots\ldots\ldots\ldots\ldots(1),$$

where a is the velocity of sound in air.

Since the motion is periodic, we may assume that $\phi = \phi' \epsilon^{int}$, whence if

$$n/a = 2\pi/\lambda = \kappa \ldots\ldots\ldots\ldots\ldots\ldots(2).$$

(1) becomes
$$\frac{d^2\phi'}{dx^2} + \kappa^2\phi' = 0,$$
the solution of which is
$$\phi' = (A \cos \kappa x + B \sin \kappa x).$$

If the pipe is closed at both ends, $d\phi/dx = 0$ when $x = 0$ and $x = l$; and since
$$\frac{d\phi}{dx} = \kappa (B \cos \kappa x - A \sin \kappa x) \, \epsilon^{\iota n t},$$
the first condition gives $B = 0$, whilst the second condition gives
$$\sin \kappa l = 0,$$
which requires that $\kappa = i\pi/l,$
where i is an integer. The value of ϕ in real quantities therefore becomes
$$\phi = A \cos i\pi x/l \cos nt \quad \dots\dots\dots\dots\dots\dots\dots(3).$$

The wave length and frequency are thus given by the equations
$$\lambda = 2l/i, \quad n/2\pi = ia/2l \quad \dots\dots\dots\dots\dots (4).$$

These equations determine the wave lengths and frequencies of the notes, which can be produced by a pipe of length l, both of whose ends are closed. The frequency of the gravest note is $a/2l$, and its wave length is $2l$; the frequencies and wave lengths of the overtones are obtained by putting $i = 2, 3\dots$

157. From (3) we see that $d\phi/dx$ vanishes whenever $x = rl/i$, where r is any integer not greater than i. Corresponding to the ith harmonic, there are therefore $i-1$ nodes which divide the pipe into i equal parts.

The increment of the pressure due to the wave motion is given by the equation
$$\delta p = -\rho \dot{\phi},$$
and therefore δp vanishes whenever $\cos i\pi x/l = 0$; i.e. whenever $x = (2r + 1) l/2i$, where r is zero or any positive integer less than i. Points at which there is no pressure variation are called *loops*. We thus see that corresponding to the gravest note ($i = 1, r = 0$), there is a loop at the middle point of the pipe. The loops corresponding to the overtones, occur at points $x = l/2i, 3l/2i\dots$; and consequently the loops bisect the distances between the nodes.

The conditions that a node may exist at any point of the pipe, can be secured by placing a rigid barrier across the interior of the pipe at that point. The conditions for a loop may be approximately realised, by making a communication at the point in question with the external air; and consequently it was assumed by Euler and Lagrange, that the open end of a pipe may be treated as a loop. This supposition is however only approximately true, but the error is small provided the diameter of the pipe is small in comparison with the wave length. Whenever a disturbance is excited in a pipe which communicates with the air, the external air is set in motion, and a complete solution of the problem would necessitate the motion of the latter being taken into account.

158. Let us in the next place suppose that one end of the pipe is fitted with a disc, which is constrained to vibrate with a velocity $\cos nt$.

The condition to be satisfied at the origin, where the disc is situated, is

$$\frac{d\phi}{dx} = \cos nt, \text{ when } x = 0.$$

If therefore we assume

$$\phi = (A \cos \kappa x + B \sin \kappa x) \cos nt,$$

we obtain $B\kappa = 1$.

If the other end of the pipe is closed, $d\phi/dx = 0$ when $x = l$, whence

$$\phi = \frac{\cos \kappa (l - x)}{\kappa \sin \kappa l} \cos nt.$$

If the other end be open, the condition is that $\phi = 0$ when $x = l$, whence

$$\phi = -\frac{\sin \kappa (l - x)}{\kappa \cos \kappa l} \cos nt.$$

The value of κ is of course n/a.

Reflection and Refraction[1].

159. We shall now investigate the reflection and refraction of plane waves of sound at the surface of separation of two gases.

Let the origin O be in the surface of separation, let the axis of x be drawn into the first medium, and let the axis of z be parallel

[1] Green, *Trans. Camb. Phil. Soc.* 1838.

REFLECTION AND REFRACTION.

to the line of intersection of the wave fronts with the surface of separation.

Let i be the angle of incidence, r the angle of refraction; also let V, V_1, be the velocities of propagation in the two gases, and ρ, ρ_1 their densities when undisturbed. Then in the first medium we must have

$$\rho' = \rho(1+s), \quad p' = k'\rho'^\gamma = k'\rho^\gamma(1+\gamma s)$$

and in the second medium

$$\rho'_1 = \rho_1(1+s_1), \quad p'_1 = k'_1\rho'^\gamma_1 = k'_1\rho^\gamma_1(1+\gamma s_1).$$

Since the two gases are supposed to be in equilibrium when undisturbed by the sound waves, we must have

$$k'\rho^\gamma = k'_1\rho^\gamma_1 \quad\quad\quad\quad\quad\quad (5).$$

Again $\quad\quad V^2 = k'\gamma\rho^{\gamma-1}, \quad V_1^2 = k'_1\gamma\rho_1^{\gamma-1},$

whence $\quad\quad V^2\rho = V_1^2\rho_1. \quad\quad\quad\quad\quad\quad (6).$

The equations of motion in the first medium are

$$\frac{d^2\phi}{dt^2} = V^2\left(\frac{d^2\phi}{dx^2} + \frac{d^2\phi}{dy^2}\right) \quad\quad\quad\quad (7),$$

$$\frac{d\phi}{dt} + V^2 s = 0 \quad\quad\quad\quad\quad\quad (8),$$

and in the second

$$\frac{d^2\phi_1}{dt^2} = V_1^2\left(\frac{d^2\phi_1}{dx^2} + \frac{d^2\phi_1}{dy^2}\right) \quad\quad\quad\quad (9),$$

$$\frac{d\phi_1}{dt} + V_1^2 s_1 = 0 \quad\quad\quad\quad\quad\quad (10).$$

The boundary conditions are,

(i) That the component velocity perpendicular to the surface of separation should be the same in both media.

(ii) That the pressure in the two media should be equal at their surface.

The first condition gives

$$\frac{d\phi}{dx} = \frac{d\phi_1}{dx} \quad\quad\quad\quad\quad\quad (11),$$

and the second gives $\quad p = p_1,$

which by (5), (8) and (10) gives

$$V_1^2\phi = V^2\phi_1. \quad\quad\quad\quad\quad\quad (12).$$

If we suppose that the velocity potential of the incident wave is

$$\phi = A e^{i(ax+by+\omega t)} \quad \text{...............(13),}$$

the velocity potentials of the reflected and the refracted waves may be written

$$\phi' = A' e^{i(a'x+by+\omega t)} \quad \text{...............(14),}$$

$$\phi_1 = A_1 e^{i(a_1 x+by+\omega t)} \quad \text{...............(15),}$$

for the coefficient of t must be the same in these three equations, because the periods $2\pi/\omega$ of the three waves must be the same; whilst the coefficients of y must be the same, because the traces of the three waves on the surface of separation must move together.

Substituting the value of $\phi + \phi'$ in (7), and the value of ϕ_1 in (9), we obtain

$$\omega^2 = V^2(a^2 + b^2) = V^2(a'^2 + b^2) = V_1^2(a_1^2 + b^2) \text{........(16),}$$

and therefore $a' = -a$. Also if λ, λ_1 be the wave lengths in the two media

$$\left. \begin{array}{l} a = (2\pi/\lambda) \cos i, \ b = (2\pi/\lambda) \sin i = (2\pi/\lambda_1) \sin r \\ a_1 = (2\pi/\lambda_1) \cos r, \ \omega = 2\pi V/\lambda = 2\pi V_1/\lambda_1 \end{array} \right\} \text{...(17).}$$

From the equation $a' = -a$, we see that the angle of incidence is equal to the angle of reflection; and from (17) it follows that

$$\frac{V}{\sin i} = \frac{V_1}{\sin r} \quad \text{...............(18),}$$

which is the law of sines.

To obtain the ratio of the amplitudes, we must substitute the values of $\phi + \phi'$ and ϕ_1 from (13), (14) and (15) in (11) and (12); we thus obtain

$$\left. \begin{array}{l} (A - A') a = A_1 a_1 \\ (A + A') V_1^2 = A_1 V^2 \end{array} \right\} \text{...............(19).}$$

By (17) and (18) these become

$$\left. \begin{array}{l} (A - A') \tan r = A_1 \tan i \\ (A + A') \sin^2 r = A_1 \sin^2 i \end{array} \right\} \text{...............(20),}$$

from which we deduce

$$A' = \frac{A \tan(i-r)}{\tan(i+r)} \quad \text{...............(21).}$$

$$A_1 = \frac{2A \sin^2 r \cot i}{\sin(i+r) \sin(i-r)} \quad \text{...............(22).}$$

The first formula is the same as Fresnel's tangent formula for the intensity of the reflected light, when the incident light is polarized perpendicularly to the plane of incidence; and we observe that the reflected wave vanishes when $i + r = \tfrac{1}{2}\pi$, i.e. when $i = \tan^{-1} V/V_1$.

160. When light is reflected at the surface of a medium, which propagates optical waves with a velocity which is greater than that of the medium from which the light proceeds, it is well known that the light will be totally reflected, when the angle of incidence exceeds a certain value which is called the *critical* angle; and that total reflection is accompanied with a change of phase. We shall now show that a similar phenomenon occurs in the case of sound.

Since
$$\cos r = \{1 - (V_1/V)^2 \sin^2 i\}^{\tfrac{1}{2}},$$
it follows that if $V_1 > V$, $\cos r$ will vanish when $i = \sin^{-1} V/V_1$, and for angles of incidence greater than this value, $\cos r$ will become imaginary; and therefore by (17), a_1 will become a negative imaginary quantity.

When $\cos r$ is imaginary, the values of A' and A_1 given by (21) and (22) become complex, and the formulae apparently fail. The explanation of this is, that the incident, reflected and refracted waves are the *real* parts of (13), (14) and (15); if therefore A' and A_1 are real, the reflected and refracted waves are given by $A' \cos(-ax + by + \omega t)$ and $A_1 \cos(a_1 x + by + \omega t)$; but if A' is complex, we must put $A' = \alpha + \iota\beta$, and the reflected wave, which is the real part of $(\alpha + \iota\beta)\, e^{\iota(-ax+by+\omega t)}$, is

$$\alpha \cos(-ax + by + \omega t) - \beta \sin(-ax + by + \omega t)$$
$$= (\alpha^2 + \beta^2)^{\tfrac{1}{2}} \cos(-ax + by + \omega t + \tan^{-1}\beta/\alpha),$$

which shows that there is a change of phase.

In order to calculate the change of phase, we must put
$$A' = \alpha + \iota\beta, \quad A_1 = \alpha_1 + \iota\beta_1, \quad \mu = V_1/V;$$
also let
$$q = (\mu^2 \sin^2 i - 1)^{\tfrac{1}{2}}/\mu \cos i.$$

From (17) we obtain
$$\frac{a_1}{a} = \frac{\lambda \cos r}{\lambda_1 \cos i} = \frac{V \cos r}{V_1 \cos i} = -\iota q,$$

whence (19) become
$$A - \alpha - \iota\beta = -\iota q(\alpha_1 + \iota\beta_1),$$
$$(A + \alpha + \iota\beta)\mu^2 = \alpha_1 + \iota\beta_1.$$

Equating the real and imaginary parts we obtain
$$\left.\begin{array}{l} A - \alpha = q\beta_1, \quad \beta = q\alpha_1 \\ (A + \alpha)\mu^2 = \alpha_1, \quad \beta\mu^2 = \beta_1 \end{array}\right\} \dots\dots\dots\dots(23),$$
whence
$$\alpha = \frac{A(1-\mu^4 q^2)}{1+\mu^4 q^2}, \quad \beta = \frac{2A\mu^2 q}{1+\mu^4 q^2}, \quad \beta_1 = \frac{2A\mu^4 q}{1+\mu^4 q^2},$$

from which we see that
$$\alpha^2 + \beta^2 = A^2,$$
$$\beta/\alpha = \tan 2e,$$
where
$$\tan e = \mu^2 q = \mu(\mu^2 \tan^2 i - \sec^2 i)^{\frac{1}{2}} \dots\dots\dots\dots(24).$$

The reflected wave is therefore
$$\phi' = A\cos(-ax + by + \omega t + 2e),$$
which shows that total reflection takes place, accompanied by a change of phase, whose value is determined by (24).

Since $\quad a_1 = -\iota q a,$
the refracted wave is
$$\phi' = (\alpha_1^2 + \beta_1^2)^{\frac{1}{2}} \epsilon^{qax} \cos(by + \omega t + \tan^{-1}\beta_1/\alpha_1).$$
where
$$qa = (2\pi/\lambda)(\sin^2 i - \mu^{-2})^{\frac{1}{2}}.$$

Since in the second medium x is negative, it follows that the refracted wave is insensible at a distance of a few wave lengths, and thus the refracted sound rapidly becomes stifled.

Spherical Waves[1].

161. We have already shown that the velocity potential satisfies the equation
$$\frac{d^2\phi}{dt^2} = a^2 \nabla^2 \phi,$$
where a is the velocity of sound; and if we assume that $\phi = \Phi\epsilon^{\iota\kappa a t}$, this becomes
$$(\nabla^2 + \kappa^2)\Phi = 0 \dots\dots\dots\dots(25).$$

---

[1] The remainder of this Chapter is taken from Lord Rayleigh's *Theory of Sound*, Vol. II. Chapter XVII. His original investigations are given in the *Proc. Lond. Math. Soc.* Vol. IV. pp. 93 and 253.

By (11) of § 7, it follows that if r, θ, ω be polar coordinates, the value of ∇^2 is

$$\nabla^2 = \frac{d^2}{dr^2} + \frac{2}{r}\frac{d}{dr} + \frac{1}{r^2 \sin\theta}\frac{d}{d\theta}\left(\sin\theta \frac{d}{d\theta}\right) + \frac{1}{r^2 \sin^2\theta}\frac{d^2}{d\omega^2};$$

if therefore the motion be symmetrical about the origin, so that Φ is a function of r alone, (25) becomes

$$\frac{d^2\Phi}{dr^2} + \frac{2}{r}\frac{d\Phi}{dr} + \kappa^2\Phi = 0,$$

which may be written in the form

$$\frac{d^2}{dr^2}(r\Phi) + \kappa^2(r\Phi) = 0,$$

the integral of which is

$$\Phi = r^{-1}(Ae^{\iota\kappa r} + Be^{-\iota\kappa r}) \quad \ldots\ldots\ldots\ldots\ldots\ldots(26).$$

If the motion is finite at the origin, we must have $A = -B$, in which case

$$\phi = 2\iota A r^{-1} e^{\iota\kappa at} \sin\kappa r \quad \ldots\ldots\ldots\ldots\ldots(27),$$

in which A may be complex.

162. This equation may be applied to determine the symmetrical vibrations of a gas, which is enclosed within a rigid spherical envelop of radius c; for the condition to be satisfied at the surface of the envelop is

$$d\phi/dr = 0,$$

which gives $\quad\quad \kappa \cos\kappa c - c^{-1}\sin\kappa c = 0,$

or $\quad\quad\quad\quad\quad \tan\kappa c = \kappa c \quad \ldots\ldots\ldots\ldots\ldots\ldots(28).$

Since the wave length $\lambda = 2\pi/\kappa$, and the frequency is equal to $\kappa a/2\pi$, (28) determines the notes which can be produced. The roots of (28) have been investigated by Lord Rayleigh, and he finds that the first root is $\kappa c = 1\cdot 4303 \times \pi$. We therefore see that the frequency of the gravest note is $\cdot 7151 \times (a/c)$; accordingly the pitch falls as the radius of the sphere increases. This result exemplifies a general law, *that the frequencies of vibration of similar bodies formed of similar materials, are inversely proportional to their linear dimensions.*

The loops are determined by the equation $\sin\kappa r = 0$, which gives $r = m\pi/\kappa$, where m is an integer.

163. Since any circular cone whose vertex is the origin is a nodal cone, the above solution determines the notes which could be produced by a conical pipe closed by a spherical segment of radius c.

If a conical pipe be open at one end, and we assume that the condition to be satisfied at the open end is that it should be a loop, we obtain $\kappa = m\pi/c$, and therefore the value of ϕ is

$$\phi = 2\iota A r^{-1} \epsilon^{\iota m \pi a t/c} \sin m\pi r/c.$$

The frequency of the gravest note is therefore $\tfrac{1}{2}a/c$, which is less than if the pipe were closed.

164. The most general value of ϕ in the case of symmetrical waves is

$$\phi = A r^{-1} \epsilon^{\iota \kappa (at+r)} + B r^{-1} \epsilon^{\iota \kappa (at-r)} \quad \ldots\ldots\ldots\ldots (29),$$

the first term of which represents waves converging upon the origin, whilst the second represents waves diverging from the origin.

Let us now draw a very small sphere surrounding the origin; then taking the second term of (29), the flux across the sphere is

$$\iint r^2 \frac{d\phi}{dr} d\Omega = - B \iint (1 + \iota \kappa r) \, \epsilon^{\iota \kappa (at-r)} d\Omega$$
$$= - 4\pi B \epsilon^{\iota \kappa at},$$

when $r = 0$. The second term of (29) therefore represents a source of sound diverging from the pole, of strength $- 4\pi B \epsilon^{\iota \kappa at}$; similarly the first term represents a source of sound converging towards the pole[1].

165. The general solution of (25) cannot be effected without the aid of spherical harmonic analysis, but there is one solution of considerable utility, which we shall now consider.

Let $\phi = \Phi \epsilon^{\iota \kappa at} \cos \theta,$

where Φ is a function of r alone. Substituting in (25), we obtain

$$\frac{d^2\Phi}{dr^2} + \frac{2}{r} \frac{d\Phi}{dr} - \frac{2}{r^2} \Phi + \kappa^2 \Phi = 0.$$

---

[1] The corresponding problems in two-dimensional motion, cannot be investigated without employing the Bessel's function of the second kind $Y_0(\kappa r)$. It is worth noticing, that certain expressions for these functions in the forms of series and definite integrals, can be obtained by means of the theory of sources of sound. See Lord Rayleigh, *Proc. Lond. Math. Soc.*, Vol. XIX. p. 504.

To solve this equation, put $\Phi = dw/dr$ and integrate; we at once obtain

$$\frac{d^2w}{dr^2} + \frac{2}{r}\frac{dw}{dr} + \kappa^2 w = 0,$$

the solution of which has already been shown to be

$$w = r^{-1}(A e^{\iota\kappa r} + B e^{-\iota\kappa r});$$

accordingly

$$\Phi = \frac{\iota\kappa}{r}(A e^{\iota\kappa r} - B e^{-\iota\kappa r}) - \frac{1}{r^2}(A e^{\iota\kappa r} + B e^{-\iota\kappa r}) \quad \ldots\ldots(30).$$

In order to find the condition that the motion should be finite at the origin, we must expand the exponentials in powers of $\iota\kappa r$, and equate the coefficients of negative powers of r to zero; we shall thus find that $A = -B$, whence writing A for $2\iota\kappa^2 A$, the solution becomes

$$\Phi = \frac{A}{\kappa r}\left(\cos\kappa r - \frac{\sin\kappa r}{\kappa r}\right) \quad \ldots\ldots\ldots\ldots\ldots(31).$$

If gas, contained in a spherical envelop, be vibrating in this manner, the frequency is determined by the equation

$$d\Phi/dr = 0, \text{ when } r = c;$$

which gives $\quad\tan\kappa c = \dfrac{2\kappa c}{2 - \kappa^2 c^2}.$

The least root of this equation (other than zero), is found by Lord Rayleigh to be $\kappa c = \cdot 662 \times \pi$; and therefore the frequency of the gravest note is $\cdot 331 \times (a/c)$.

This note is the gravest note which can be produced by gas vibrating within a sphere; it is more than an octave lower than the gravest radial vibration, whose frequency has been shown to be $\cdot 7151 \times (a/c)$.

Since the motion is symmetrical with respect to the diameter $\theta = 0$, every meridional plane is a nodal plane; but since $d\Phi/d\theta$ does not vanish anywhere except along the diameter in question, there are no conical nodal sheets.

166. We shall now consider the motion of a spherical pendulum surrounded with air, which is performing small oscillations.

Since the periods of the pendulum and of the air must be the same, we may suppose the velocity of the pendulum to be represented by $Ve^{\iota\kappa at}$, and therefore the condition to be satisfied at the surface of the sphere is

$$d\phi/dr = Ve^{\iota\kappa at}\cos\theta \quad \ldots\ldots\ldots\ldots\ldots\ldots(32).$$

The form of this equation suggests that ϕ must vary as $\cos\theta$; we shall therefore assume that $\phi = \Phi e^{\iota\kappa at}\cos\theta$, where Φ is given by (30). Since the disturbance is propagated outwards, $A = 0$, and therefore

$$\Phi = -Br^{-2}(1 + \iota\kappa r)\,\epsilon^{-\iota\kappa r}.$$

Substituting in (32), we obtain

$$B = \frac{Vc^3 \epsilon^{\iota\kappa c}}{2 - \kappa^2 c^2 + 2\iota\kappa c} \quad \ldots\ldots\ldots\ldots\ldots\ldots(33),$$

where c is the radius of the sphere.

If X be the resistance experienced by the sphere,

$$X = \iint \delta p \cos\theta dS$$
$$= -\iint \rho\dot\phi \cos\theta dS$$
$$= -\tfrac{4}{3}\pi\rho c^2 \iota\kappa a \Phi e^{\iota\kappa at}$$
$$= \tfrac{4}{3}\pi\rho c^3 a\dot\xi \frac{\iota\kappa(1+\iota\kappa c)}{2 - \kappa^2 c^2 + 2\iota\kappa c},$$

where $\xi = Ve^{\iota\kappa at}$, is the velocity of the sphere.

Rationalising the denominator, and putting

$$p = \frac{2 + \kappa^2 c^2}{4 + \kappa^4 c^4}, \quad q = \frac{\kappa^3 c^3}{4 + \kappa^4 c^4},$$

and remembering that $\dot\xi = \iota\kappa a\xi$, we obtain

$$X = M'(p\dot\xi + \kappa a q\xi),$$

where M' is the mass of the displaced fluid.

The first term of this expression represents an increase in the inertia of the sphere; whilst the second term represents a resistance proportional to the velocity, which is therefore a *viscous* term, and shows that initial energy is gradually dissipated into space. If M be the mass of the sphere, l the distance of its centre from the point of suspension, the equation of motion of the pendulum is

$$\{M(l^2 + \tfrac{2}{5}c^2) + M'lp\}\,\ddot\theta + M'l^2\kappa a q\dot\theta + (M - M')gl\theta = 0.$$

By § 122, the integral of this equation is of the form
$$\theta = A\epsilon^{-\delta t} \sin(\mu t + \alpha),$$
and the modulus of decay is
$$2\{M(l^2 + \tfrac{2}{5}c^2) + M'l^2 p\}/M'l^2 \kappa a q.$$

If the wave length λ, of the vibrations of the gas, is large in comparison with the radius of the sphere, κc will be of the order c/λ, and will therefore be small; accordingly the value of p will be nearly equal to $\tfrac{1}{2}$, whilst the value of κq, upon which the viscous term depends, will be of the order c^2/λ^4. We therefore see that in this case the viscous term will be very small, and the motion will die away gradually; hence the sphere will vibrate very nearly in the same manner as if the gas were an incompressible fluid.

If, on the other hand, c were large compared with λ, p would be nearly equal to unity, and the apparent inertia of the sphere would be greater than when c/λ is small; but κq would be of the order c^{-1} and would therefore be small.

167. Another interesting problem is that of the scattering of a plane wave of sound by a fixed rigid sphere, whose diameter is small compared with the wave length.

Measuring θ from the direction of propagation, the velocity potential of the plane waves may be taken to be
$$\phi = \epsilon^{\iota\kappa(at+x)} = \epsilon^{\iota\kappa(at+r\cos\theta)},$$
the positive sign being taken, because the waves are supposed to be travelling in the negative direction of the axis of x.

If c be the radius of the sphere, it follows that in the neighbourhood of the sphere, κr or $2\pi r/\lambda$ is a small quantity, and therefore expanding the exponential and dropping the time factor for the present, we may arrange ϕ in the form of the series[1]
$$\phi = 1 - \tfrac{1}{6}\kappa^2 r^2 + \iota\kappa r \cos\theta - \tfrac{1}{6}\kappa^2 r^2 (3\cos^2\theta - 1)\ldots$$

When the waves impinge upon the sphere, a reflected or

---

[1] The reader, who is acquainted with Spherical Harmonic analysis, will observe that we have arranged ϕ in a series of zonal harmonics. It can be shown that the solution of (25) can be expressed in a series of terms of the type $F(r) S_n$, where S_n is a spherical surface harmonic.

scattered wave is thrown off, whose velocity potential may be assumed to be

$$\phi' = A_0\Phi_0 + A_1\Phi_1\cos\theta + \tfrac{1}{2}A_2\Phi_2(3\cos^2\theta - 1) + \ldots$$

The quantities Φ_0, Φ_1 are given by (26) and (30) respectively; but since the scattered wave diverges from the sphere, we must put $A = 0$, and take $B = 1$, since the constant B may be supposed to be included in $A_0, A_1, \ldots$; accordingly

$$\left.\begin{array}{l}\Phi_0 = r^{-1}e^{-\iota\kappa r}\\ \Phi_1 = -r^{-2}(1 + \iota\kappa r)e^{-\iota\kappa r}\end{array}\right\}\quad\ldots\ldots\ldots\ldots(35).$$

With regard to Φ_2, it can be verified by trial, that a solution of (25) is $\Phi_2(3\cos^2\theta - 1)$, where Φ_2 is a function of r alone; it will not however be necessary to consider the form of Φ_2, since it introduces quantities of a higher order than $\kappa^2 c^2$, which will be neglected.

The equation to be satisfied at the surface of the sphere is

$$\frac{d\phi}{dr} + \frac{d\phi'}{dr} = 0,$$

when $r = c$. This equation must hold good for all values of θ, whence

$$A_0\frac{d\Phi_0}{dr} - \tfrac{1}{3}\kappa^2 c = 0,$$

$$A_1\frac{d\Phi_1}{dr} + \iota\kappa = 0,$$

which determine A_0, A_1. Substituting from (35) we obtain

$$A_0 = -\frac{\kappa^2 c^3 e^{\iota\kappa c}}{3(1 + \iota\kappa c)} = -\tfrac{1}{3}\kappa^2 c^3,$$

$$A_1 = -\frac{\iota\kappa c^3 e^{\iota\kappa c}}{2 + 2\iota\kappa c - \kappa^2 c^2} = -\tfrac{1}{2}\iota\kappa c^3$$

approximately, since we shall not retain powers of c higher than c^3. We thus obtain

$$\phi' = -\frac{e^{-\iota\kappa r}}{r}\left(\tfrac{1}{3}\kappa^2 c^3 - \frac{1 + \iota\kappa r}{2r}\iota\kappa c^3\cos\theta\right).$$

At a considerable distance from the sphere, the term $\kappa c^3/r^2$ is so small that it may be neglected, we may therefore write

$$\phi' = -\frac{e^{-\iota\kappa r}}{3r}(1 + \tfrac{3}{2}\cos\theta)\kappa^2 c^3.$$

Restoring the time factor and putting $\kappa = 2\pi/\lambda$, we finally obtain in real quantities

$$\phi' = -\frac{4\pi^2 c^3}{3\lambda^2 r}(1 + \tfrac{3}{2}\cos\theta)\cos\frac{2\pi}{\lambda}(at - r)\ldots\ldots\ldots(36),$$

corresponding to the wave

$$\phi = \cos\frac{2\pi}{\lambda}(at + x)\ldots\ldots\ldots\ldots\ldots(37).$$

Equation (36) accordingly gives the velocity potential of the scattered wave, corresponding to the incident wave whose velocity potential is given by (37). This expression is however only an approximate one, and the correctness of the approximation depends upon the assumption, that the radius of the sphere is so small in comparison with the wave length, that terms of a higher order than c^3/λ^2 may be neglected. We have also neglected $\kappa c^3/r^2$, which is equivalent to supposing, that the point at which we are observing the effect of the scattered wave, is at a considerable distance from the sphere. For a more complete investigation, we must refer to Lord Rayleigh's treatise.

EXAMPLES.

1. If two simple tones of equal intensity and having a given small difference of pitch be heard together, prove that the number of beats in a given time will be greater, the higher the two simple tones are in the musical scale; and prove that the pitch of the resultant sound in the course of each beat is constant.

2. One end of a tube which contains air is open, whilst the other is fitted with a disc, which vibrates in such a manner that the pressure of the air in contact with the disc is

$$\Pi(1 - k\sin 2\pi t/\tau)$$

where k is a small quantity. Find the velocity potential of the motion.

3. The radius of a solid sphere surrounded by an unlimited mass of air, is given by $R(1 + \alpha\sin nat)$, where a is the velocity of sound in air. Show that the mean energy per unit of mass

of air at a distance r from the centre of the sphere, due to the motion of the latter is

$$\tfrac{1}{3} n^2 a^2 R^6 (1 + 2n^2 r^2)/r^4 (1 + n^2 R^2).$$

4. Prove that in order that indefinite plane waves may be transmitted without alteration, with uniform velocity a in a homogeneous fluid medium, the pressure and density must be connected by the equation

$$p - p_0 = a^2 \rho_0^2 (\rho_0^{-1} - \rho^{-1}),$$

where p_0, ρ_0 are the pressure and density in the undisturbed part of the fluid.

5. Two gases of densities ρ, ρ_1 are separated by a plane uniform flexible membrane, whose equation is $y = 0$, and whose superficial density and tension are σ and T. If plane waves of sound impinge obliquely at an angle i, and the displacements of the incident reflected and refracted waves of sound and of the membrane, be represented by

(i) $A \sin \{m (x \sin i - y \cos i) - nt + \alpha\}$,

(ii) $A' \sin \{m (x \sin i + y \cos i) - nt + \alpha'\}$,

(iii) $A_1 \sin \{m_1 (x \sin r - y \cos r) - nt + \alpha_1\}$,

(iv) $a \sin (mx \sin i - nt)$,

respectively; find the relations to be satisfied, and prove that the ratio of the intensities of the reflected and incident waves is equal to

$$\frac{(Tm^2 \sin^2 i - \sigma n^2)^2 + (\rho_1 m_1 \sec r - \rho m \sec i)^2}{(Tm^2 \sin^2 i - \sigma n^2)^2 + (\rho m \sec r + \rho m \sec i)^2}.$$

6. If sound waves be travelling along a straight tube of infinite length which is adiathermanous, and no conduction of heat takes place through the air, prove that the equations of motion may be accurately satisfied by supposing a wave of condensation to travel along the tube, with a velocity of propagation which at each point depends only on the condensation at that point, and which for a density ρ is

$$\left[1 + \frac{\gamma+1}{\gamma-1} \left\{ \left(\frac{\rho}{\rho_0}\right)^{\frac{1}{2}(\gamma-1)} - 1 \right\} \right] \sqrt{\frac{p_0 \gamma}{\rho_0}},$$

where p_0, ρ_0 are the pressure and density at each end of the wave.

7. Prove that in a closed endless uniform tube of length l filled with air, a piston of mass M will perform m complete small vibrations under the elasticity of a spring, if

$$\tan m\pi l/a = \frac{Mm\pi l}{M'a}\left(\frac{n^2}{m^2} - 1\right),$$

where M' is the mass of the air in the tube, and a the velocity of sound, supposing the piston to make n vibrations in a second when the air is exhausted.

8. Investigate the forced oscillations in a straight pipe, which will occur when the temperature of air in the pipe is compelled to undergo small harmonic vibrations expressed by $\theta \cos m(vt - x)$, where x is measured along the axis of the pipe.

9. The greatest angle inclination of the adiabatic lines of a gas to its isothermals occurs, when the slope of the isothermal to the line of zero pressure is $\pi - \cot^{-1}\gamma$; and the locus of all these points of maximum angle, is a straight line through the origin, inclined to the line of zero pressure at an angle $\cot^{-1}\gamma^{\frac{1}{2}}$.

10. A sphere of mean radius R, executes simple harmonic radial vibrations of amplitude α, in air of density ρ; prove that its energy is radiated into the atmosphere in sound waves at the rate

$$2\pi\rho a\left(\frac{a\alpha}{\lambda}\right)^2 \frac{(2\pi R)^4}{(2\pi R)^2 + \lambda^2}$$

per unit of time, where λ is the length of the waves propagated in air, and a is their velocity.

NOTE TO § 53.

The proposition at the end of § 53 is not quite accurately stated, inasmuch as the constraint contemplated must be equivalent to an increase in the inertia of the system. When this is the case, the periods of vibration are increased and consequently the frequency is diminished. The proposition is not however true, when the constraint is not of this character. For example, in § 163 we have shown that the frequency of the gravest note of an open conical pipe of length c is equal to $\frac{1}{2}a/c$, whilst by § 162 the frequency of the gravest note of a closed conical pipe is $\cdot 7151 \times (a/c)$; and therefore in this case, the effect of constraint is to diminish the period and increase the frequency.

A

CLASSIFIED CATALOGUE

OF

EDUCATIONAL WORKS

PUBLISHED BY

GEORGE BELL & SONS

LONDON: YORK STREET, COVENT GARDEN
NEW YORK: 66, FIFTH AVENUE; AND BOMBAY
CAMBRIDGE: DEIGHTON, BELL & CO

December, 1894

CONTENTS.

	PAGE
GREEK AND LATIN CLASSICS:—	
Annotated and Critical Editions	3
Texts	9
Translations	10
Grammar and Composition	15
History, Geography, and Reference Books, etc.	18
MATHEMATICS:—	
Arithmetic and Algebra	19
Bookkeeping	20
Geometry and Euclid	21
Analytical Geometry, etc.	21
Trigonometry	22
Mechanics and Natural Philosophy	22
MODERN LANGUAGES:—	
English	24
French Class Books	29
French Annotated Editions	31
German Class Books	31
German Annotated Editions	32
Italian	34
Bell's Modern Translations	34
SCIENCE, TECHNOLOGY AND ART:—	
Chemistry	34
Botany	35
Geology	35
Medicine	36
Bell's Agricultural Series	36
Technological Handbooks	37
Music	37
Art	38
MENTAL, MORAL AND SOCIAL SCIENCES:—	
Psychology and Ethics	39
History of Philosophy	40
Law and Political Economy	40
History	41
Divinity, etc.	42
Summary of Series	45

GREEK AND LATIN CLASSICS.

ANNOTATED AND CRITICAL EDITIONS.

AESCHYLUS. Edited by F. A. PALEY. M.A., LL.D., late Classical Examiner to the University of London. *4th edition, revised.* 8vo, 8s.
[*Bib. Class.*
— Edited by F. A. PALEY, M.A., LL.D., 6 vols. fcap. 8vo, 1s. 6d.
[*Camb. Texts with Notes.*

Agamemnon.	Persae.
Choephoroe.	Prometheus Vinctus.
Eumenides.	Septem contra Thebas.

ARISTOPHANIS Comoediae quae supersunt cum perditarum fragmentis tertiis curis, recognovit additis adnotatione critica, summariis, descriptione metrica, onomastico lexico HUBERTUS A. HOLDEN, LL.D. [late Fellow of Trinity College, Cambridge]. Demy 8vo.

Vol. I., containing the Text expurgated, with Summaries and Critical Notes, 18s.

The Plays sold separately:

Acharnenses, 2s.	Aves, 2s.
Equites, 1s. 6d.	Lysistrata, et Thesmophoriazusae, 4s.
Nubes, 2s.	
Vespae, 2s.	Ranae, 2s.
Pax, 2s.	Plutus, 2s.

Vol. II. Onomasticon Aristophaneum continens indicem geographicum et historicum   5s. 6d.

— The Peace. A revised Text with English Notes and a Preface. By F. A. PALEY, M.A., LL.D.   Post 8vo, 4s. 6d.   [*Pub. Sch. Ser.*
— The Acharnians. A revised Text with English Notes and a Preface. By F. A. PALEY, M.A., LL.D.   Post 8vo, 4s. 6d.   [*Pub. Sch. Ser.*
— The Frogs. A revised Text with English Notes and a Preface. By F. A. PALEY, M.A., LL.D.   Post 8vo, 4s. 6d.   [*Pub. Sch. Ser.*

CAESAR De Bello Gallico. Edited by GEORGE LONG, M.A. *New edition.* Fcap. 8vo, 4s.

Or in parts, Books I.-III., 1s. 6d.; Books IV. and V., 1s. 6d.; Books VI. and VII., 1s. 6d. |*Gram. Sch. Class.*

— De Bello Gallico.   Book I.   Edited by GEORGE LONG, M.A. With Vocabulary by W. F. R. SHILLETO, M.A. 1s. 6d.   [*Lower Form Ser.*
— De Bello Gallico.   Book II.   Edited by GEORGE LONG, M.A. With Vocabulary by W. F. R. SHILLETO, M.A. Fcap. 8vo, 1s. 6d.
[*Lower Form Ser.*
— De Bello Gallico. Book III. Edited by GEORGE LONG, M.A. With Vocabulary by W. F. R. SHILLETO, M.A. Fcap. 8vo, 1s. 6d.
[*Lower Form Ser.*
— Seventh Campaign in Gaul. B.C. 52. De Bello Gallico, Lib. VII. Edited with Notes, Excursus, and Table of Idioms, by REV. W. COOKWORTHY COMPTON, M.A., Head Master of Dover College. With Illustrations from Sketches by E. T. COMPTON, Maps and Plans. *2nd edition.*   Crown 8vo, 2s. 6d. net.

" A really admirable class book."—*Spectator.*

" One of the most original and interesting books which have been published in late years as aids to the study of classical literature. I think

CAESAR De Bello Gallico—*continued.*
it gives the student a new idea of the way in which a classical book may be made a living reality."—*Rev. J. E. C. Welldon*, Harrow.
— **Easy Selections from the Helvetian War.** Edited by A. M. M. STEDMAN, M.A. With Introduction, Notes and Vocabulary. 18mo. 1*s.*
[*Primary Classics.*
CALPURNIUS SICULUS and M. AURELIUS OLYMPIUS NEMESIANUS. The Eclogues, with Introduction, Commentary, and Appendix. By C. H. KEENE, M.A. Crown 8vo, 6*s.*
CATULLUS, TIBULLUS, and PROPERTIUS. Selected Poems. Edited by the REV. A. H. WRATISLAW, late Head Master of Bury St. Edmunds School, and F. N. SUTTON, B.A. With Biographical Notices of the Poets. Fcap. 8vo, 2*s.* 6*d.* [*Gram. Sch. Class.*
CICERO'S Orations. Edited by G. LONG, M.A. 8vo. [*Bib. Class.*
   Vol. I.—In Verrem. 8*s.*
   Vol. II.—Pro P. Quintio—Pro Sex. Roscio—Pro. Q. Roscio—Pro M. Tullio—Pro M. Fonteio—Pro A. Caecina—De Imperio Cn. Pompeii—Pro A. Cluentio—De Lege Agraria—Pro C. Rabirio. 8*s.*
   Vols. III. and IV. *Out of print.*
— **De Senectute, De Amicitia, and Select Epistles.** Edited by GEORGE LONG, M.A. *New edition.* Fcap. 8vo, 3*s.* [*Gram. Sch. Class.*
— **De Amicitia.** Edited by GEORGE LONG, M.A. Fcap. 8vo, 1*s.* 6*d.*
[*Camb. Texts with Notes.*
— **De Senectute.** Edited by GEORGE LONG, M.A. Fcap. 8vo, 1*s.* 6*d.*
[*Camb. Texts with Notes.*
— **Epistolae Selectae.** Edited by GEORGE LONG, M.A. Fcap. 8vo, 1*s.* 6*d.*
[*Camb. Texts with Notes.*
— **The Letters to Atticus.** Book I. With Notes, and an Essay on the Character of the Writer. By A. PRETOR, M.A., late of Trinity College, Fellow of St. Catherine's College, Cambridge. *3rd edition.* Post 8vo, 4*s.* 6*d.* [*Pub. Sch. Ser.*
CORNELIUS NEPOS. Edited by the late REV. J. F. MACMICHAEL, Head Master of the Grammar School, Ripon. Fcap. 8vo, 2*s.*
[*Gram. Sch. Class.*
DEMOSTHENES. Edited by R. WHISTON, M.A., late Head Master of Rochester Grammar School. 2 vols. 8vo, 8*s.* each. [*Bib. Class.*
   Vol. I.—Olynthiacs—Philippics—De Pace—Halonnesus—Chersonese —Letter of Philip—Duties of the State—Symmoriae—Rhodians—Megalopolitans—Treaty with Alexander—Crown.
   Vol. II.—Embassy—Leptines—Meidias—Androtion—Aristocrates—Timocrates—Aristogeiton.
— **De Falsa Legatione.** By the late R. SHILLETO, M.A., Fellow of St. Peter's College, Cambridge. *7th edition.* Post 8vo, 6*s.* [*Pub. Sch. Ser.*
— **The Oration against the Law of Leptines.** With English Notes. By the late B. W. BEATSON, M.A., Fellow of Pembroke College. *3rd edition.* Post 8vo, 3*s.* 6*d.* [*Pub. Sch. Ser.*
EURIPIDES. By F. A. PALEY, M.A., LL.D. 3 vols. *2nd edition, revised.* 8vo, 8*s.* each. Vol. I. *Out of print.* [*Bib. Class.*
   Vol. II.—Preface—Ion—Helena—Andromache—Electra—Bacchae—Hecuba. 2 Indexes.
   Vol. III.—Preface—Hercules Furens—Phoenissae—Orestes—Iphigenia in Tauris—Iphigenia in Aulide—Cyclops. 2 Indexes.

EURIPIDES. Electra. Edited, with Introduction and Notes, by C. H. KEENE, M.A., Dublin, Ex-Scholar and Gold Medallist in Classics. Demy 8vo, 10s. 6d.
— Edited by F. A. PALEY, M.A., LL.D. 13 vols. Fcap. 8vo, 1s. 6d. each.
[Camb. Texts with Notes.

   Alcestis.   Phoenissae.
   Medea.   Troades.
   Hippolytus.   Hercules Furens.
   Hecuba.   Andromache.
   Bacchae.   Iphigenia in Tauris.
   Ion (2s.).   Supplices.
   Orestes.

HERODOTUS. Edited by REV. J. W. BLAKESLEY, B.D. 2 vols. 8vo, 12s.
[Bib. Class.
— Easy Selections from the Persian Wars. Edited by A. G. LIDDELL, M.A. With Introduction, Notes, and Vocabulary. 18mo, 1s. 6d.
[Primary Classics.

HESIOD. Edited by F. A. PALEY, M.A., LL.D. *2nd edition, revised.* 8vo, 5s.
[Bib. Class.

HOMER. Edited by F. A. PALEY, M.A., LL.D. 2 vols. *2nd edition, revised.* 14s. Vol. II. (Books 13-24) may be had separately, 6s.
[Bib. Class.
— Iliad. Books I.-XII. Edited by F. A. PALEY, M.A., LL.D. Fcap. 8vo, 4s. 6d.
   Also in 2 Parts. Books I.-VI. 2s. 6d. Books VII.-XII. 2s. 6d.
[Gram. Sch. Class.
— Iliad. Book I. Edited by F. A. PALEY, M.A., LL.D. Fcap. 8vo, 1s.
[Camb. Text with Notes.

HORACE. Edited by REV. A. J. MACLEANE, M.A. *4th edition,* revised by GEORGE LONG. 8vo, 8s. *[Bib. Class.*
— Edited by A. J. MACLEANE, M.A. With a short Life. Fcap. 8vo, 3s. 6d.
   Or, Part I., Odes, Carmen Seculare, and Epodes, 2s.; Part II., Satires, Epistles, and Art of Poetry, 2s. *[Gram. Sch. Class.*
— Odes. Book I. Edited by A. J. MACLEANE, M.A. With a Vocabulary by A. H. DENNIS, M.A. Fcap. 8vo, 1s. 6d. *[Lower Form Ser.*

JUVENAL: Sixteen Satires (expurgated). By HERMAN PRIOR, M.A., late Scholar of Trinity College, Oxford. Fcap. 8vo, 3s. 6d.
[Gram. Sch. Class.

LIVY. The first five Books, with English Notes. By J. PRENDEVILLE. A new edition revised throughout, and the notes in great part re-written, by J. H. FREESE, M.A., late Fellow of St. John's College, Cambridge. Books I. II. III. IV. V. With Maps and Introductions. Fcap. 8vo. 1s. 6d. each.
— Book VI. Edited by E. S. WEYMOUTH, M.A., Lond., and G. F. HAMILTON, B.A. With Historical Introduction, Life of Livy, Notes, Examination Questions, Dictionary of Proper Names, and Map. Crown 8vo, 2s. 6d.
— Book XXI. By the REV. L. D. DOWDALL, M.A., late Scholar and University Student of Trinity College, Dublin, B.D., Ch. Ch. Oxon. Post 8vo, 3s. 6d. *[Pub. Sch. Ser.*
— Book XXII. Edited by the REV. L. D. DOWDALL, M.A., B.D. Post 8vo, 3s. 6d. *[Pub. Sch. Ser.*

LIVY. Easy Selections from the Kings of Rome. Edited by A. M. M. STEDMAN, M.A. With Introduction, Notes, and Vocabulary. 18mo, 1s. 6d. [*Primary Class.*

LUCAN. The Pharsalia. By C. E. HASKINS, M.A., Fellow of St. John's College, Cambridge, with an Introduction by W. E. HEITLAND, M.A., Fellow and Tutor of St. John's College, Cambridge. 8vo, 14s.

LUCRETIUS. Titi Lucreti Cari De Rerum Natura Libri Sex. By the late H. A. J. MUNRO, M.A., Fellow of Trinity College, Cambridge. *4th edition, finally revised.* 3 vols, demy 8vo. Vols. I., II., Introduction, Text, and Notes, 18s. Vol. III., Translation, 6s.

MARTIAL: Select Epigrams. Edited by F. A. PALEY, M.A., LL.D., and the late W. H. STONE, Scholar of Trinity College, Cambridge. With a Life of the Poet. Fcap. 8vo, 4s. 6d. [*Gram. Sch. Class.*

OVID: Fasti. Edited by F. A. PALEY, M.A., LL.D. *Second edition.* Fcap. 8vo, 3s. 6d. [*Gram. Sch. Class.*
Or in 3 vols, 1s. 6d. each [*Grammar School Classics*], or 2s. each [*Camb. Texts with Notes*], Books I. and II., Books III. and IV., Books V. and VI.

— Selections from the Amores, Tristia, Heroides, and Metamorphoses. By A. J. MACLEANE, M.A. Fcap. 8vo, 1s. 6d. [*Camb. Texts with Notes.*

— Ars Amatoria et Amores. A School Edition. Carefully Revised and Edited, with some Literary Notes, by J. HERBERT WILLIAMS, M.A., late Demy of Magdalen College, Oxford. Fcap. 8vo, 3s. 6d.

— Heroides XIV. Edited, with Introductory Preface and English Notes, by ARTHUR PALMER, M.A., Professor of Latin at Trinity College, Dublin. Demy 8vo, 6s.

— Metamorphoses, Book XIII. A School Edition. With Introduction and Notes, by CHARLES HAINES KEENE, M.A., Dublin, Ex-Scholar and Gold Medallist in Classics. *3rd edition.* Fcap. 8vo, 2s. 6d.

— Epistolarum ex Ponto Liber Primus. With Introduction and Notes, by CHARLES HAINES KEENE, M.A. Crown 8vo, 3s.

PLATO. The Apology of Socrates and Crito. With Notes, critical and exegetical, by WILHELM WAGNER, PH.D. *12th edition.* Post 8vo, 3s. 6d. A CHEAP EDITION. Limp Cloth. 2s. 6d. [*Pub. Sch. Ser.*

— Phaedo. With Notes, critical and exegetical, and an Analysis, by WILHELM WAGNER, PH.D. *9th edition.* Post 8vo, 5s. 6d. [*Pub. Sch. Ser.*

— Protagoras. The Greek Text revised, with an Analysis and English Notes, by W. WAYTE, M.A., Classical Examiner at University College, London. *7th edition.* Post 8vo, 4s. 6d. [*Pub. Sch. Ser.*

— Euthyphro. With Notes and Introduction by G. H. WELLS, M.A., Scholar of St. John's College, Oxford; Assistant Master at Merchant Taylors' School. *3rd edition.* Post 8vo, 3s. [*Pub. Sch. Ser.*

— The Republic. Books I. and II. With Notes and Introduction by G. H. WELLS, M.A. *4th edition*, with the Introduction re-written. Post 8vo, 5s. [*Pub. Sch. Ser.*

— Euthydemus. With Notes and Introduction by G. H. WELLS, M.A. Post 8vo, 4s. [*Pub. Sch. Ser.*

— Phaedrus. By the late W. H. THOMPSON, D.D., Master of Trinity College, Cambridge. 8vo, 5s. [*Bib. Class.*

— Gorgias. By the late W. H. THOMPSON, D.D., Master of Trinity College, Cambridge. *New edition.* 6s. [*Pub. Sch. Ser.*

PLAUTUS. Aulularia. With Notes, critical and exegetical, by W. WAGNER, PH.D. *5th edition.* Post 8vo, 4s. 6d. [*Pub. Sch. Ser.*
— **Trinummus.** With Notes, critical and exegetical, by WILHELM WAGNER, PH.D. *5th edition.* Post 8vo, 4s. 6d. [*Pub. Sch. Ser.*
— **Menaechmei.** With Notes, critical and exegetical, by WILHELM WAGNER, PH.D. *2nd edition.* Post 8vo, 4s. 6d. [*Pub. Sch. Ser.*
— **Mostellaria.** By E. A. SONNENSCHEIN, M.A., Professor of Classics at Mason College, Birmingham. Post 8vo, 5s. [*Pub. Sch. Ser.*
— **Captivi.** Abridged and Edited for the Use of Schools. With Introduction and Notes by J. H. FREESE, M.A., formerly Fellow of St. John's College, Cambridge. Fcap. 8vo, 1s. 6d.
PROPERTIUS. Sex. Aurelii Propertii Carmina. The Elegies of Propertius, with English Notes. By F. A. PALEY, M.A., LL.D. *2nd edition.* 8vo, 5s.
SALLUST : Catilina and Jugurtha. Edited, with Notes, by the late GEORGE LONG. *New edition, revised*, with the addition of the Chief Fragments of the Histories, by J. G. FRAZER, M.A., Fellow of Trinity College, Cambridge. Fcap. 8vo, 3s. 6d, or separately, 2s. each.
[*Gram. Sch. Class.*
SOPHOCLES. Edited by REV. F. H. BLAYDES, M.A. Vol. I. Oedipus Tyrannus—Oedipus Coloneus—Antigone. 8vo, 8s. [*Bib. Class.*
Vol. II. Philoctetes—Electra—Trachiniae—Ajax. By F. A. PALEY, M.A., LL.D. 8vo, 6s., or the four Plays separately in limp cloth, 2s. 6d. each.
— **Trachiniae.** With Notes and Prolegomena. By ALFRED PRETOR, M.A., Fellow of St. Catherine's College, Cambridge. Post 8vo, 4s. 6d.
[*Pub. Sch. Ser.*
— **The Oedipus Tyrannus of Sophocles.** By B. H. KENNEDY, D.D., Regius Professor of Greek and Hon. Fellow of St. John's College, Cambridge. With a Commentary containing a large number of Notes selected from the MS. of the late T. H. STEEL, M.A. Crown 8vo, 8s.
— — A SCHOOL EDITION, post 8vo, 5s. [*Pub. Sch. Ser.*
— Edited by F. A. PALEY, M.A., LL.D. 5 vols. Fcap. 8vo, 1s. 6d. each.
[*Camb. Texts with Notes.*

Oedipus Tyrannus.	Electra.
Oedipus Coloneus.	Ajax.
Antigone.	

TACITUS : Germania and Agricola. Edited by the late REV. P. FROST, late Fellow of St. John's College, Cambridge. Fcap. 8vo, 2s. 6d.
[*Gram. Sch. Class.*
— **The Germania.** Edited, with Introduction and Notes, by R. F. DAVIS, M.A. Fcap. 8vo, 1s. 6d.
TERENCE. With Notes, critical and explanatory, by WILHELM WAGNER, PH.D. *3rd edition.* Post 8vo, 7s. 6d. [*Pub. Sch. Ser.*
— Edited by WILHELM WAGNER, PH.D. 4 vols. Fcap. 8vo, 1s. 6d. each.
[*Camb. Texts with Notes.*

| Andria. | Hautontimorumenos. |
| Adelphi. | Phormio. |

THEOCRITUS. With short, critical and explanatory Latin Notes, by F. A. PALEY, M.A., LL.D. *2nd edition, revised.* Post 8vo, 4s. 6d.
[*Pub. Sch. Ser.*

THUCYDIDES, Book VI. By T. W. DOUGAN, M.A., Fellow of St. John's College, Cambridge; Professor of Latin in Queen's College, Belfast. Edited with English notes. Post 8vo, 3*s.* 6*d.* [*Pub. Sch. Ser.*

— The History of the Peloponnesian War. With Notes and a careful Collation of the two Cambridge Manuscripts, and of the Aldine and Juntine Editions. By the late RICHARD SHILLETO, M.A., Fellow of St. Peter's College, Cambridge. 8vo. Book I. 6*s.* 6*d.* Book II. 5*s.* 6*d.*

VIRGIL. By the late PROFESSOR CONINGTON, M.A. Revised by the late PROFESSOR NETTLESHIP, Corpus Professor of Latin at Oxford. 8vo.
[*Bib. Class.*

    Vol. I. The Bucolics and Georgics, with new Memoir and three Essays on Virgil's Commentators, Text, and Critics. *4th edition.* 10*s.* 6*d.*
    Vol. II. The Aeneid, Books I.-VI. *4th edition.* 10*s.* 6*d.*
    Vol. III. The Aeneid, Books VII.-XII. *3rd edition.* 10*s.* 6*d.*

— Abridged from PROFESSOR CONINGTON'S Edition, by the REV. J. G. SHEPPARD, D.C.L., H. NETTLESHIP, late Corpus Professor of Latin at the University of Oxford, and W. WAGNER, PH.D. 2 vols. fcap. 8vo, 4*s.* 6*d.* each. [*Gram. Sch. Class.*

    Vol. I. Bucolics, Georgics, and Aeneid, Books I.-IV.
    Vol. II. Aeneid, Books V.-XII.
    Also the Bucolics and Georgics, in one vol. 3*s.*

Or in 9 separate volumes (Grammar School Classics, with Notes at foot of page), price 1s. 6d. each.

Bucolics.	Aeneid, V. and VI.
Georgics, I. and II.	Aeneid, VII. and VIII.
Georgics, III. and IV.	Aeneid, IX. and X.
Aeneid, I. and II.	Aeneid, XI. and XII.
Aeneid, III. and IV.	

Or in 12 separate volumes (Cambridge Texts with Notes at end), price 1s. 6d. each.

Bucolics.	Aeneid, VII.
Georgics, I. and II.	Aeneid, VIII.
Georgics, III. and IV.	Aeneid, IX.
Aeneid, I. and II.	Aeneid, X.
Aeneid, III. and IV.	Aeneid, XI.
Aeneid, V. and VI. (price 2*s.*)	Aeneid, XII.

— Aeneid, Book I. CONINGTON'S Edition abridged. With Vocabulary by W F. R. SHILLETO, M.A. Fcap. 8vo, 1*s.* 6*d.* [*Lower Form Ser.*

XENOPHON : Anabasis. With Life, Itinerary, Index, and three Maps. Edited by the late J. F. MACMICHAEL. *Revised edition.* Fcap. 8vo, 3*s.* 6*d.* [*Gram. Sch. Class.*

    Or in 4 separate volumes, price 1s. 6d. each.

    Book I. (with Life, Introduction, Itinerary, and three Maps)—Books II. and III.—Books IV. and V.—Books VI. and VII.

— **Anabasis.** MACMICHAEL'S Edition, revised by J. E. MELHUISH, M.A., Assistant Master of St. Paul's School. In 6 volumes, fcap. 8vo. With Life, Itinerary, and Map to each volume, 1*s.* 6*d.* each.
[*Camb. Texts with Notes.*

    Book I.—Books II. and III.—Book IV.—Book V.—Book VI.—Book VII.

Educational Catalogue. 9

XENOPHON. Cyropaedia. Edited by G. M. GORHAM, M.A., late Fellow of Trinity College, Cambridge. *New edition.* Fcap. 8vo, 3s. 6d.
[*Gram. Sch. Class.*
 Also Books I. and II., 1s. 6d.; Books V. and VI., 1s. 6d.
— Memorabilia. Edited by PERCIVAL FROST, M.A., late Fellow of St. John's College, Cambridge. Fcap. 8vo, 3s. [*Gram. Sch. Class.*
— Hellenica. Book I. Edited by L. D. DOWDALL, M.A., B.D. Fcap. 8vo, 2s. [*Camb. Texts with Notes.*
— Hellenica. Book II. By L. D. DOWDALL, M.A., B.D. Fcap. 8vo, 2s.
[*Camb. Texts with Notes.*

TEXTS.

AESCHYLUS. Ex novissima recensione F. A. PALEY, A.M., LL.D. Fcap. 8vo, 2s. [*Camb. Texts.*
CAESAR De Bello Gallico. Recognovit G. LONG, A.M. Fcap. 8vo, 1s. 6d. [*Camb. Texts.*
CATULLUS. A New Text, with Critical Notes and an Introduction, by J. P. POSTGATE, M.A., LITT.D., Fellow of Trinity College, Cambridge, Professor of Comparative Philology at the University of London. Wide fcap. 8vo, 3s.
CICERO De Senectute et de Amicitia, et Epistolae Selectae. Recensuit G. LONG, A.M. Fcap. 8vo, 1s. 6d. [*Camb. Texts.*
CICERONIS Orationes in Verrem. Ex recensione G. LONG, A.M. Fcap. 8vo, 2s. 6d. [*Camb. Texts.*
CORPUS POETARUM LATINORUM, a se aliisque denuo recognitorum et brevi lectionum varietate instructorum, edidit JOHANNES PERCIVAL POSTGATE. Tom. I.—Ennius, Lucretius, Catullus, Horatius, Vergilius, Tibullus, Propertius, Ovidius. Large post 4to, 21s. net. Also in 2 Parts, sewed, 9s. each, net.
 ⁎⁎ To be completed in 4 parts, making 2 volumes.
CORPUS POETARUM LATINORUM. Edited by WALKER. Containing :—Catullus, Lucretius, Virgilius, Tibullus, Propertius, Ovidius, Horatius, Phaedrus, Lucanus, Persius, Juvenalis, Martialis, Sulpicia, Statius, Silius Italicus, Valerius Flaccus, Calpurnius Siculus, Ausonius, and Claudianus. 1 vol. 8vo, cloth, 18s.
EURIPIDES. Ex recensione F. A. PALEY, A.M., LL.D. 3 vols. Fcap. 8vo, 2s. each. [*Camb. Texts.*
 Vol. I.—Rhesus—Medea— Hippolytus — Alcestis —Heraclidae—Supplices—Troades.
 Vol. II.—Ion—Helena—Andromache—Electra—Bacchae—Hecuba.
 Vol. III.—Hercules Furens—Phoenissae—Orestes—Iphigenia in Tauris —Iphigenia in Aulide—Cyclops.
HERODOTUS. Recensuit J. G. BLAKESLEY, S.T.B. 2 vols. Fcap. 8vo, 2s. 6d. each. [*Camb. Texts.*
HOMERI ILIAS I.-XII. Ex novissima recensione F. A. PALEY, A.M., LL.D. Fcap. 8vo, 1s. 6d. [*Camb. Texts.*
HORATIUS. Ex recensione A. J. MACLEANE. A.M. Fcap. 8vo, 1s. 6d.
[*Camb. Texts.*
JUVENAL ET PERSIUS. Ex recensione A. J. MACLEANE, A.M. Fcap. 8vo, 1s. 6d. [*Camb. Texts.*

LUCRETIUS. Recognovit H. A. J. MUNRO, A.M. Fcap. 8vo, 2s.
[*Camb. Texts.*]
PROPERTIUS. Sex. Propertii Elegiarum Libri IV. recensuit A. PALMER, collegii sacrosanctae et individuae Trinitatis juxta Dublinum Socius. Fcap. 8vo, 3s. 6d.
SALLUSTI CRISPI CATILINA ET JUGURTHA, Recognovit G. LONG, A.M. Fcap. 8vo, 1s. 6d. [*Camb. Texts.*]
SOPHOCLES. Ex recensione F. A. PALEY, A.M., LL.D. Fcap. 8vo, 2s. 6d.
[*Camb. Texts.*]
TERENTI COMOEDIAE. GUL. WAGNER relegit et emendavit. Fcap. 8vo, 2s. [*Camb. Texts.*]
THUCYDIDES. Recensuit J. G. DONALDSON, S.T.P. 2 vols. Fcap. 8vo, 2s. each. [*Camb. Texts.*]
VERGILIUS. Ex recensione J. CONINGTON, A.M. Fcap. 8vo, 2s.
[*Camb. Texts.*]
XENOPHONTIS EXPEDITIO CYRI. Recensuit J. F. MACMICHAEL. A.B. Fcap. 8vo, 1s. 6d. [*Camb. Texts.*]

TRANSLATIONS.

AESCHYLUS, The Tragedies of. Translated into English Prose. By F. A. PALEY, M.A., LL.D., Editor of the Greek Text. *2nd edition revised*, 8vo, 7s. 6d.
— **The Tragedies of.** Translated into English verse by ANNA SWANWICK. *4th edition revised*. Small post 8vo, 5s.
— **The Tragedies of.** Literally translated into Prose, by T. A. BUCKLEY, B.A. Small post 8vo, 3s. 6d.
— **The Tragedies of.** Translated by WALTER HEADLAM, M.A., Fellow of King's College, Cambridge. [*Preparing.*]
ANTONINUS (M. Aurelius), The Thoughts of. Translated by GEORGE LONG, M.A. *Revised edition.* Small post 8vo, 3s. 6d.
Fine paper edition on handmade paper. Pott 8vo, 6s.
APOLLONIUS RHODIUS. The Argonautica. Translated by E. P. COLERIDGE. Small post 8vo, 5s.
AMMIANUS MARCELLINUS. History of Rome during the Reigns of Constantius, Julian, Jovianus, Valentinian, and Valens. Translated by PROF. C. D. YONGE, M.A. With a complete Index. Small post 8vo, 7s. 6d.
ARISTOPHANES, The Comedies of. Literally translated by W. J. HICKIE. *With Portrait.* 2 vols. small post 8vo, 5s. each.
Vol. I.—Acharnians, Knights, Clouds, Wasps, Peace, and Birds.
Vol. II.—Lysistrata, Thesmophoriazusae, Frogs, Ecclesiazusae, and Plutus.
— **The Acharnians.** Translated by W. H. COVINGTON, B.A. With Memoir and Introduction. Crown 8vo, sewed, 1s.
ARISTOTLE on the Athenian Constitution. Translated, with Notes and Introduction, by F. G. KENYON, M.A., Fellow of Magdalen College, Oxford. Pott 8vo, printed on handmade paper. *2nd edition.* 4s. 6d.
— **History of Animals.** Translated by RICHARD CRESSWELL, M.A. Small post 8vo, 5s.

ARISTOTLE. Organon: or, Logical Treatises, and the Introduction of Porphyry. With Notes, Analysis, Introduction, and Index, by the REV. O. F. OWEN, M.A. 2 vols. small post 8vo, 3s. 6d. each.
— **Rhetoric and Poetics.** Literally Translated, with Hobbes' Analysis, &c., by T. BUCKLEY, B.A. Small post 8vo, 5s.
— **Nicomachean Ethics.** Literally Translated, with Notes, an Analytical Introduction, &c., by the Venerable ARCHDEACON BROWNE, late Classical Professor of King's College. Small post 8vo, 5s.
— **Politics and Economics.** Translated, with Notes, Analyses, and Index, by E. WALFORD, M.A., and an Introductory Essay and a Life by DR. GILLIES. Small post 8vo, 5s.
— **Metaphysics.** Literally Translated, with Notes, Analysis, &c., by the REV. JOHN H. M'MAHON, M.A. Small post 8vo, 5s.
ARRIAN. Anabasis of Alexander, together with the Indica. Translated by E. J. CHINNOCK, M.A., LL.D. With Introduction, Notes, Maps, and Plans. Small post 8vo, 5s.
CAESAR. Commentaries on the Gallic and Civil Wars, with the Supplementary Books attributed to Hirtius, including the complete Alexandrian, African, and Spanish Wars. Translated by W. A. M'DEVITTE, B.A. Small post 8vo, 5s.
— **Gallic War.** Translated by W. A. M'DEVITTE, B.A. 2 vols., with Memoir and Map. Crown 8vo, sewed. Books I. to IV., Books V. to VII., 1s. each.
CALPURNIUS SICULUS, The Eclogues of. The Latin Text, with English Translation by E. J. L. SCOTT, M.A. Crown 8vo, 3s. 6d.
CATULLUS, TIBULLUS, and the Vigil of Venus. Prose Translation. Small post 8vo, 5s.
CICERO, The Orations of. Translated by PROF. C. D. YONGE, M.A. With Index. 4 vols. small post 8vo, 5s. each.
— **On Oratory and Orators.** With Letters to Quintus and Brutus. Translated by the REV. J. S. WATSON, M.A. Small post 8vo, 5s.
— **On the Nature of the Gods.** Divination, Fate, Laws, a Republic, Consulship. Translated by PROF. C. D YONGE, M.A., and FRANCIS BARHAM. Small post 8vo, 5s.
— **Academics, De Finibus, and Tusculan Questions.** By PROF. C. D. YONGE, M.A. Small post 8vo, 5s.
— **Offices;** or, Moral Duties. Cato Major, an Essay on Old Age; Laelius, an Essay on Friendship; Scipio's Dream; Paradoxes; Letter to Quintus on Magistrates. Translated by C. R. EDMONDS. *With Portrait,* 3s. 6d
— **Old Age and Friendship.** Translated, with Memoir and Notes, by G. H. WELLS, M.A. Crown 8vo, sewed, 1s.
DEMOSTHENES, The Orations of. Translated, with Notes, Arguments, a Chronological Abstract, Appendices, and Index, by C. RANN KENNEDY. 5 vols. small post 8vo.
Vol. I.—The Olynthiacs, Philippics. 3s. 6d.
Vol. II.—On the Crown and on the Embassy. 5s.
Vol. III.—Against Leptines, Midias, Androtion, and Aristocrates. 5s.
Vols. IV. and V.—Private and Miscellaneous Orations. 5s. each.
— **On the Crown.** Translated by C. RANN KENNEDY. Small post 8vo, sewed, 1s., cloth, 1s 6d.
DIOGENES LAERTIUS. Translated by PROF. C. D. YONGE, M.A. Small post 8vo, 5s.

EPICTETUS, The Discourses of. With the Encheiridion and Fragments. Translated by GEORGE LONG, M.A. Small post 8vo, 5s.
Fine Paper Edition, 2 vols. Pott 8vo, 10s. 6d.

EURIPIDES. A Prose Translation, from the Text of Paley. By E. P. COLERIDGE, B.A. 2 vols., 5s. each.
Vol. I.—Rhesus, Medea, Hippolytus, Alcestis, Heraclidæ, Supplices, Troades, Ion, Helena.
Vol. II.—Andromache, Electra, Bacchae, Hecuba, Hercules Furens, Phoenissae, Orestes, Iphigenia in Tauris, Iphigenia in Aulis, Cyclops.
*** The plays separately (except Rhesus, Helena, Electra, Iphigenia in Aulis, and Cyclops). Crown 8vo, sewed, 1s. each.
— Translated from the Text of Dindorf. By T. A. BUCKLEY, B.A. 2 vols. small post 8vo, 5s. each.

GREEK ANTHOLOGY. Translated by GEORGE BURGES, M.A. Small post 8vo, 5s.

HERODOTUS. Translated by the REV. HENRY CARY, M.A. Small post 8vo, 3s. 6d.
— Analysis and Summary of. By J. T. WHEELER. Small post 8vo, 5s.

HESIOD, CALLIMACHUS, and THEOGNIS. Translated by the REV. J. BANKS, M.A. Small post 8vo, 5s.

HOMER. The Iliad. Translated by T. A. BUCKLEY, B.A. Small post 8vo, 5s.
— The Odyssey, Hymns, Epigrams, and Battle of the Frogs and Mice. Translated by T. A. BUCKLEY, B.A. Small post 8vo, 5s.
— The Iliad. Books I.-IV. Translated into English Hexameter Verse, by HENRY SMITH WRIGHT, B.A., late Scholar of Trinity College, Cambridge. Medium 8vo, 5s.

HORACE. Translated by Smart. *Revised edition.* By T. A. BUCKLEY, B.A. Small post 8vo, 3s. 6d.
— The Odes and Carmen Saeculare. Translated into English Verse by the late JOHN CONINGTON, M.A., Corpus Professor of Latin in the University of Oxford. 11th edition. Fcap. 8vo. 3s. 6d.
— The Satires and Epistles. Translated into English Verse by PROF. JOHN CONINGTON, M.A. 8th edition. Fcap. 8vo, 3s. 6d.
— Odes and Epodes. Translated by SIR STEPHEN E. DE VERE, BART. 3rd edition, enlarged. Imperial 16mo. 7s. 6d. net.

ISOCRATES, The Orations of. Translated by J. H. FREESE, M.A., late Fellow of St. John's College, Cambridge, with Introductions and Notes. Vol. I. Small post 8vo, 5s.

JUSTIN, CORNELIUS NEPOS, and EUTROPIUS. Translated by the REV. J. S. WATSON, M.A. Small post 8vo, 5s.

JUVENAL, PERSIUS, SULPICIA, and LUCILIUS. Translated by L. EVANS, M.A. Small post 8vo, 5s.

LIVY. The History of Rome. Translated by DR. SPILLAN, C. EDMONDS, and others. 4 vols. small post 8vo, 5s. each.
— Books I., II., III., IV. A Revised Translation by J. H. FREESE, M.A., late Fellow of St. John's College, Cambridge. With Memoir, and Maps. 4 vols., crown 8vo, sewed, 1s. each.
— Book V. A Revised Translation by E. S. WEYMOUTH, M.A., Lond. With Memoir, and Maps. Crown 8vo, sewed, 1s.
— Book IX. Translated by FRANCIS STORR, B.A. With Memoir. Crown 8vo, sewed, 1s.

Educational Catalogue. 13

LUCAN. The Pharsalia. Translated into Prose by H. T. RILEY. Small post 8vo, 5s.
— The Pharsalia. Book I. Translated by FREDERICK CONWAY, M.A. With Memoir and Introduction. Crown 8vo, sewed, 1s.
LUCIAN'S Dialogues of the Gods, of the Sea-Gods, and of the Dead. Translated by HOWARD WILLIAMS, M.A. Small post 8vo, 5s.
LUCRETIUS. Translated by the REV. J. S. WATSON, M.A. Small post 8vo, 5s.
— Literally translated by the late H. A. J. MUNRO, M.A. *4th edition.* Demy 8vo, 6s.
MARTIAL'S Epigrams, complete. Literally translated into Prose, with the addition of Verse Translations selected from the Works of English Poets, and other sources. Small post 8vo, 7s. 6d.
OVID, The Works of. Translated. 3 vols., small post 8vo, 5s. each.
　Vol. I.— Fasti, Tristia, Pontic Epistles, Ibis, and Halieuticon.
　Vol. II.—Metamorphoses. *With Frontispiece.*
　Vol. III.—Heroides, Amours, Art of Love, Remedy of Love, and Minor Pieces. *With Frontispiece.*
PINDAR. Translated by DAWSON W. TURNER. Small post 8vo, 5s.
PLATO. Gorgias. Translated by the late E. M. COPE, M.A., Fellow of Trinity College. *2nd edition.* 8vo, 7s.
— Philebus. Translated by F. A. PALEY, M.A., LL.D. Small 8vo, 4s.
— Theaetetus. Translated by F. A. PALEY, M.A., LL.D. Small 8vo, 4s.
— The Works of. Translated, with Introduction and Notes. 6 vols. small post 8vo, 5s. each.
　Vol. I.—The Apology of Socrates—Crito—Phaedo—Gorgias—Protagoras—Phaedrus—Theaetetus—Eutyphron—Lysis. Translated by the REV. H. CARY.
　Vol. II.—The Republic—Timaeus—Critias. Translated by HENRY DAVIS.
　Vol. III.—Meno—Euthydemus—The Sophist—Statesman—Cratylus—Parmenides—The Banquet. Translated by G. BURGES.
　Vol. IV.—Philebus—Charmides—Laches—Menexenus—Hippias—Ion—The Two Alcibiades—Theages—Rivals—Hipparchus—Minos—Clitopho—Epistles. Translated by G. BURGES.
　Vol. V.—The Laws. Translated by G. BURGES.
　Vol. VI.—The Doubtful Works. Edited by G. BURGES. With General Index to the six volumes.
— Apology, Crito, Phaedo, and Protagoras. Translated by the REV. H. CARY. Small post 8vo, sewed, 1s., cloth, 1s. 6d.
— Dialogues. A Summary and Analysis of. With Analytical Index, giving references to the Greek text of modern editions and to the above translations. By A. DAY, LL.D. Small post 8vo, 5s.
PLAUTUS, The Comedies of. Translated by H. T. RILEY, B.A. 2 vols. small post 8vo, 5s. each.
　Vol. I.—Trinummus—Miles Gloriosus—Bacchides—Stichus—Pseudolus—Menaechmei—Aulularia—Captivi—Asinaria—Curculio.
　Vol. II.—Amphitryon—Rudens—Mercator—Cistellaria—Truculentus—Persa—Casina—Poenulus—Epidicus—Mostellaria—Fragments.
— Trinummus, Menaechmei, Aulularia, and Captivi. Translated by H. T. RILEY, B.A. Small post 8vo, sewed, 1s., cloth, 1s. 6d.

PLINY. The Letters of Pliny the Younger. Melmoth's Translation, revised, by the REV. F. C. T. BOSANQUET, M.A. Small post 8vo, 5s.

PLUTARCH. Lives. Translated by A. STEWART, M.A., late Fellow of Trinity College, Cambridge, and GEORGE LONG, M.A. 4 vols. small post 8vo, 3s. 6d. each.

— **Morals.** Theosophical Essays. Translated by C. W. KING, M.A., late Fellow of Trinity College, Cambridge. Small post 8vo, 5s.

— **Morals.** Ethical Essays. Translated by the REV. A. R. SHILLETO, M.A. Small post 8vo, 5s.

PROPERTIUS. Translated by REV. P. J. F. GANTILLON, M.A., and accompanied by Poetical Versions, from various sources. Small post 8vo, 3s. 6d.

PRUDENTIUS, Translations from. A Selection from his Works, with a Translation into English Verse, and an Introduction and Notes, by FRANCIS ST. JOHN THACKERAY, M.A., F.S.A., Vicar of Mapledurham, formerly Fellow of Lincoln College, Oxford, and Assistant-Master at Eton. Wide post 8vo, 7s. 6d.

QUINTILIAN: Institutes of Oratory, or, Education of an Orator. Translated by the REV. J. S. WATSON, M.A. 2 vols. small post 8vo, 5s. each.

SALLUST, FLORUS, and **VELLEIUS PATERCULUS.** Translated by J. S. WATSON, M.A. Small post 8vo, 5s.

SENECA: On Benefits. Translated by A. STEWART, M.A., late Fellow of Trinity College, Cambridge. Small post 8vo, 3s. 6d.

— **Minor Essays** and **On Clemency.** Translated by A. STEWART, M.A. Small post 8vo, 5s.

SOPHOCLES. Translated, with Memoir, Notes, etc., by E. P. COLERIDGE, B.A. Small post 8vo, 5s.
Or the plays separately, crown 8vo, sewed, 1s. each.

— **The Tragedies of.** The Oxford Translation, with Notes, Arguments, and Introduction. Small post 8vo, 5s.

— **The Dramas of.** Rendered in English Verse, Dramatic and Lyric, by SIR GEORGE YOUNG, BART., M.A., formerly Fellow of Trinity College, Cambridge. 8vo, 12s. 6d.

— **The Œdipus Tyrannus.** Translated into English Prose. By PROF. B. H. KENNEDY. Crown 8vo, in paper wrapper, 1s.

SUETONIUS. Lives of the Twelve Caesars and Lives of the Grammarians. Thomson's revised Translation, by T. FORESTER. Small post 8vo, 5s.

TACITUS, The Works of. Translated, with Notes and Index. 2 vols. Small post 8vo, 5s. each.
Vol. I.—The Annals.
Vol. II.—The History, Germania, Agricola, Oratory, and Index.

TERENCE and **PHAEDRUS.** Translated by H. T. RILEY, B.A. Small post 8vo, 5s.

THEOCRITUS, BION, MOSCHUS, and **TYRTAEUS.** Translated by the REV. J. BANKS, M.A. Small post 8vo, 5s.

THEOCRITUS. Translated into English Verse by C. S. CALVERLEY, M.A., late Fellow of Christ's College, Cambridge. *New edition, revised.* Crown 8vo, 7s. 6d.

THUCYDIDES. The Peloponnesian War. Translated by the REV. H. DALE. *With Portrait.* 2 vols., 3s. 6d. each.
— Analysis and Summary of. By J. T. WHEELER. Small post 8vo, 5s.
VIRGIL. Translated by A. HAMILTON BRYCE, LL.D. With Memoir and Introduction. Small post 8vo, 3s. 6d.
 Also in 6 vols., crown 8vo, sewed, 1s. each.
 Georgics. | Æneid IV.-VI.
 Bucolics. | Æneid VII.-IX.
 Æneid I.-III. | Æneid X.-XII.
XENOPHON. The Works of. In 3 vols. Small post 8vo, 5s. each.
 Vol. I.—The Anabasis, and Memorabilia. Translated by the REV. J. S. WATSON, M.A. With a Geographical Commentary, by W. F. AINSWORTH, F.S.A., F.R.G.S., etc.
 Vol. II.—Cyropaedia and Hellenics. Translated by the REV. J. S. WATSON, M.A., and the REV. H. DALE.
 Vol. III.—The Minor Works. Translated by the REV. J. S. WATSON, M.A.
SABRINAE COROLLA In Hortulis Regiae Scholae Salopiensis contexuerunt tres viri floribus legendis. *4th edition, revised and re-arranged.* By the late BENJAMIN HALL KENNEDY, D.D., Regius Professor of Greek at the University of Cambridge. Large post 8vo, 10s. 6d.
SERTUM CARTHUSIANUM Floribus trium Seculorum Contextum. Cura GULIELMI HAIG BROWN, Scholae Carthusianae Archididascali. Demy 8vo, 5s.
TRANSLATIONS into English and Latin. By C. S. CALVERLEY, M.A., late Fellow of Christ's College, Cambridge. *3rd edition.* Crown 8vo, 7s. 6d.
TRANSLATIONS from and into the Latin, Greek and English. By R. C. JEBB, M.A., Regius Professor of Greek in the University of Cambridge, H. JACKSON, M.A., LITT. D., Fellows of Trinity College, Cambridge, and W. E. CURREY, M.A., formerly Fellow of Trinity College, Cambridge. Crown 8vo. *2nd edition, revised.* 8s.

GRAMMAR AND COMPOSITION.

BADDELEY. Auxilia Latina. A Series of Progressive Latin Exercises. By M. J. B. BADDELEY, M.A. Fcap. 8vo. Part I., Accidence. *5th edition.* 2s. Part II. *5th edition.* 2s. Key to Part II. 2s. 6d.
BAIRD. Greek Verbs. A Catalogue of Verbs, Irregular and Defective; their leading formations, tenses in use, and dialectic inflexions, with a copious Appendix, containing Paradigms for conjugation, Rules for formation of tenses, &c., &c. By J. S. BAIRD, T.C.D. *New edition, revised.* 2s. 6d.
— Homeric Dialect. Its Leading Forms and Peculiarities. By J. S. BAIRD, T.C.D. *New edition, revised.* By the REV. W. GUNION RUTHERFORD, M.A., LL.D., Head Master at Westminster School. 1s.
BAKER. Latin Prose for London Students. By ARTHUR BAKER, M.A., Classical Master, Independent College, Taunton. Fcap. 8vo, 2s.

BARRY. Notes on Greek Accents. By the RIGHT REV. A. BARRY, D.D. *New edition, re-written.* 1s.

CHURCH. Latin Prose Lessons. By A. J. CHURCH, M.A., Professor of Latin at University College, London. *9th edition.* Fcap. 8vo, 2s. 6d.

CLAPIN. Latin Primer. By the REV. A. C. CLAPIN, M.A., Assistant Master at Sherborne School. *3rd edition.* Fcap. 8vo, 1s.

COLLINS. Latin Exercises and Grammar Papers. By T. COLLINS, M.A., Head Master of the Latin School, Newport, Salop. *7th edition.* Fcap. 8vo, 2s. 6d.

— Unseen Papers in Latin Prose and Verse. With Examination Questions. *6th edition.* Fcap. 8vo, 2s. 6d.

— Unseen Papers in Greek Prose and Verse. With Examination Questions. *3rd edition.* Fcap. 8vo, 3s.

— Easy Translations from Nepos, Caesar, Cicero, Livy, &c., for Retranslation into Latin. With Notes. 2s.

COMPTON. Rudiments of Attic Construction and Idiom. An Introduction to Greek Syntax for Beginners who have acquired some knowledge of Latin. By the REV. W. COOKWORTHY COMPTON, M.A., Head Master of Dover College. Crown 8vo, 3s.

FROST. Eclogae Latinae; or, First Latin Reading Book. With Notes and Vocabulary by the late REV. P. FROST, M.A. Fcap. 8vo, 1s. 6d.

— Analecta Graeca Minora. With Notes and Dictionary. *New edition.* Fcap. 8vo, 2s.

— Materials for Latin Prose Composition. By the late REV. P. FROST, M.A. *New edition.* Fcap. 8vo. 2s. Key. 4s. net.

— A Latin Verse Book. *New edition.* Fcap. 8vo, 2s. Key. 5s. net.

— Materials for Greek Prose Composition. *New edition.* Fcap. 8vo, 2s. 6d. Key. 5s. net.

— Greek Accidence. *New edition.* 1s.

— Latin Accidence. 1s.

HARKNESS. A Latin Grammar. By ALBERT HARKNESS. Post 8vo, 6s.

KEY. A Latin Grammar. By the late T. H. KEY, M.A., F.R.S. *6th thousand.* Post 8vo, 8s.

— A Short Latin Grammar for Schools. *16th edition.* Post 8vo, 3s. 6d.

HOLDEN. Foliorum Silvula. Part I. Passages for Translation into Latin Elegiac and Heroic Verse. By H. A. HOLDEN, LL.D. *11th edition.* Post 8vo, 7s. 6d.

— Foliorum Silvula. Part II. Select Passages for Translation into Latin Lyric and Comic Iambic Verse. *3rd edition.* Post 8vo, 5s.

— Foliorum Centuriae. Select Passages for Translation into Latin and Greek Prose. *10th edition.* Post 8vo, 8s.

JEBB, JACKSON, and CURREY. Extracts for Translation in Greek, Latin, and English. By R. C. JEBB, LITT.D., LL.D., Regius Professor of Greek in the University of Cambridge; H. JACKSON, LITT.D., Fellow of Trinity College, Cambridge; and W. E. CURREY, M.A., late Fellow of Trinity College, Cambridge. 4s. 6d.

Latin Syntax, Principles of. 1s.

Latin Versification. 1s.

MASON. Analytical Latin Exercises By C. P. MASON, B.A. *4th edition.* Part I., 1s. 6d. Part II., 2s. 6d.

— The Analysis of Sentences Applied to Latin. Post 8vo, 1s. 6d.

NETTLESHIP. Passages for Translation into Latin Prose. Preceded by Essays on :—I. Political and Social Ideas II. Range of Metaphorical Expression. III. Historical Development of Latin Prose Style in Antiquity. IV Cautions as to Orthography. By H. NETTLESHIP, M.A., late Corpus Professor of Latin in the University of Oxford. Crown 8vo, 3s. A Key. 4s. 6d. net.

Notabilia Quaedam; or the Principal Tenses of most of the Irregular Greek Verbs, and Elementary Greek, Latin, and French Constructions. *New edition.* 1s.

PALEY. Greek Particles and their Combinations according to Attic Usage. A Short Treatise. By F. A. PALEY, M.A., LL.D. 2s. 6d.

PENROSE. Latin Elegiac Verse, Easy Exercises in. By the REV. J. PENROSE. *New edition.* 2s. (Key, 3s. 6d. net.)

PRESTON. Greek Verse Composition. By G. PRESTON, M.A. *5th edition.* Crown 8vo, 4s. 6d.

PRUEN. Latin Examination Papers. Comprising Lower, Middle, and Upper School Papers, and a number of the Woolwich and Sandhurst Standards. By G. G. PRUEN, M.A., Senior Classical Master in the Modern Department, Cheltenham College. Crown 8vo, 2s. 6d.

SEAGER. Faciliora. An Elementary Latin Book on a New Principle. By the REV. J. L. SEAGER, M.A. 2s. 6d.

STEDMAN (A. M. M.). First Latin Lessons. By A. M. M. STEDMAN, M.A., Wadham College, Oxford. *2nd edition, enlarged.* Crown 8vo, 2s.

— Initia Latina. Easy Lessons on Elementary Accidence. *2nd edition.* Fcap. 8vo, 1s.

— First Latin Reader. With Notes adapted to the Shorter Latin Primer and Vocabulary. Crown 8vo, 1s. 6d.

— Easy Latin Passages for Unseen Translation. *2nd and enlarged edition.* Fcap. 8vo, 1s. 6d.

— Exempla Latina. First Exercises in Latin Accidence. With Vocabulary. Crown 8vo, 1s. 6d.

— The Latin Compound Sentence; Rules and Exercises. Crown 8vo, 1s. 6d. With Vocabulary, 2s.

— Easy Latin Exercises on the Syntax of the Shorter and Revised Latin Primers. With Vocabulary. *3rd edition.* Crown 8vo, 2s. 6d.

— Latin Examination Papers in Miscellaneous Grammar and Idioms. *3rd edition.* 2s. 6d. Key (for Tutors only), 6s. net.

— Notanda Quaedam. Miscellaneous Latin Exercises. On Common Rules and Idioms. *2nd edition.* Fcap. 8vo 1s. 6d. With Vocabulary, 2s.

— Latin Vocabularies for Repetition. Arranged according to Subjects. *3rd edition.* Fcap. 8vo, 1s. 6d.

— First Greek Lessons. [*In preparation.*

— Easy Greek Passages for Unseen Translation. Fcap. 8vo, 1s. 6d.

— Easy Greek Exercises on Elementary Syntax. [*In preparation.*

— Greek Vocabularies for Repetition. Fcap. 8vo, 1s. 6d.

— Greek Testament Selections for the Use of Schools. *2nd edition.* With Introduction, Notes, and Vocabulary. Fcap. 8vo, 2s. 6d.

— Greek Examination Papers in Miscellaneous Grammar and Idioms. *2nd edition.* 2s. 6d. Key (for Tutors only), 6s. net.

THACKERAY. Anthologia Graeca. A Selection of Greek Poetry, with Notes. By F. ST. JOHN THACKERAY. *5th edition.* 16mo, 4s. 6d.

THACKERAY. Anthologia Latina. A Selection of Latin Poetry, from Naevius to Boëthius, with Notes. By REV. F. ST. JOHN THACKERAY. *6th edition.* 16mo, 4s. 6d.
— Hints and Cautions on Attic Greek Prose Composition. Crown 8vo, 3s. 6d.
— Exercises on the Irregular and Defective Greek Verbs. 1s. 6d.
WELLS. Tales for Latin Prose Composition. With Notes and Vocabulary. By G. H. WELLS, M.A., Assistant Master at Merchant Taylor's School. Fcap. 8vo, 2s.

HISTORY, GEOGRAPHY, AND REFERENCE BOOKS, ETC.

TEUFFEL'S History of Roman Literature. *5th edition*, revised by DR. SCHWABE, translated by PROFESSOR G. C. W. WARR, M.A, King's College, London. Medium 8vo. 2 vols. 30s. Vol. I. (The Republican Period), 15s. Vol. II. (The Imperial Period), 15s.
KEIGHTLEY'S Mythology of Ancient Greece and Italy. *4th edition*, revised by the late LEONHARD SCHMITZ, PH.D., LL.D., Classical Examiner to the University of London. With 12 Plates. Small post 8vo, 5s.
DONALDSON'S Theatre of the Greeks. *10th edition.* Small post 8vo, 5s.
DICTIONARY OF LATIN AND GREEK QUOTATIONS; including Proverbs, Maxims, Mottoes, Law Terms and Phrases. With all the Quantities marked, and English Translations. With Index Verborum. Small post 8vo, 5s.
A GUIDE TO THE CHOICE OF CLASSICAL BOOKS. By J. B. MAYOR, M.A., Professor of Moral Philosophy at King's College, late Fellow and Tutor of St. John's College, Cambridge. *3rd edition*, with Supplementary List. Crown 8vo, 4s. 6d. Supplement separate, 1s. 6d.
PAUSANIAS' Description of Greece. Newly translated, with Notes and Index, by A. R. SHILLETO, M.A. 2 vols. Small post 8vo, 5s. each.
STRABO'S Geography. Translated by W. FALCONER, M.A., and H. C. HAMILTON. 3 vols. Small post 8vo, 5s. each.
AN ATLAS OF CLASSICAL GEOGRAPHY. By W. HUGHES and G. LONG, M.A. Containing Ten selected Maps. Imp. 8vo, 3s.
AN ATLAS OF CLASSICAL GEOGRAPHY. Twenty-four Maps by W. HUGHES and GEORGE LONG, M.A. With coloured outlines. Imperial 8vo, 6s.
ATLAS OF CLASSICAL GEOGRAPHY. 22 large Coloured Maps. With a complete Index. Imp. 8vo, chiefly engraved by the Messrs. Walker. 7s. 6d.

MATHEMATICS.

ARITHMETIC AND ALGEBRA.

BARRACLOUGH (T.). The Eclipse Mental Arithmetic. By TITUS BARRACLOUGH, Board School, Halifax. Standards I., II., and III., sewed, 6*d*.; Standards II., III., and IV., sewed, 6*d*. net; Book III., Part A, sewed, 4*d*.; Book III., Part B, cloth, 1*s*. 6*d*.

BEARD (W. S.). Graduated Exercises in Addition (Simple and Compound). For Candidates for Commercial Certificates and Civil Service appointments. By W. S. BEARD, F.R.G.S., Head Master of the Modern School, Fareham. 2*nd edition*. Fcap. 4to, 1*s*.

— *See* PENDLEBURY.

ELSEE (C.). Arithmetic. By the REV. C. ELSEE, M.A., late Fellow of St. John's College, Cambridge, Senior Mathematical Master at Rugby School. 14*th edition*. Fcap. 8vo, 3*s*. 6*d*.
[*Camb. School and College Texts.*

— **Algebra.** By the REV. C. ELSEE, M.A. 8*th edition*. Fcap. 8vo, 4*s*.
[*Camb. S. and C. Texts.*

FILIPOWSKI (H. E.). Anti-Logarithms, A Table of. By H. E. FILIPOWSKI. 3*rd edition*. 8vo, 15*s*.

GOUDIE (W. P.). *See* Watson.

HATHORNTHWAITE (J. T.). Elementary Algebra for Indian Schools. By J. T. HATHORNTHWAITE, M.A., Principal and Professor of Mathematics at Elphinstone College, Bombay. Crown 8vo, 2*s*.

HUNTER (J.). Supplementary Arithmetic, with Answers. By REV. J. HUNTER, M.A. Fcap. 8vo, 3*s*.

MACMICHAEL (W. F.) and PROWDE SMITH (R.). Algebra. A Progressive Course of Examples. By the REV. W. F. MACMICHAEL, and R. PROWDE SMITH, M.A. 4*th edition*. Fcap. 8vo, 3*s*. 6*d*. With answers, 4*s*. 6*d*.
[*Camb. S. and C. Texts.*

MATHEWS (G. B.). Theory of Numbers. An account of the Theories of Congruencies and of Arithmetical Forms. By G. B. MATHEWS, M.A., Professor of Mathematics in the University College of North Wales. Part I. Demy 8vo, 12*s*.

PENDLEBURY (C.). Arithmetic. With Examination Papers and 8,000 Examples. By CHARLES PENDLEBURY, M.A., F.R.A.S., Senior Mathematical Master of St. Paul's, Author of "Lenses and Systems of Lenses, treated after the manner of Gauss." 7*th edition*. Crown 8vo. Complete, with or without Answers, 4*s*. 6*d*. In Two Parts, with or without Answers, 2*s*. 6*d*. each.
Key to Part II. 7*s*. 6*d*. net. [*Camb. Math. Ser.*

— **Examples in Arithmetic.** Extracted from Pendlebury's Arithmetic. With or without Answers. 5*th edition*. Crown 8vo, 3*s*., or in Two Parts, 1*s*. 6*d*. and 2*s*. [*Camb. Math. Ser.*

— **Examination Papers in Arithmetic.** Consisting of 140 papers, each containing 7 questions; and a collection of 357 more difficult problems. 2*nd edition*. Crown 8vo, 2*s*. 6*d*. Key, for Tutors only, 5*s*. net.

PENDLEBURY (C.) and TAIT (T. S.). Arithmetic for Indian Schools. By C. PENDLEBURY, M.A. and T. S. TAIT, M.A., B.SC., Principal of Baroda College. Crown 8vo, 3s. [*Camb. Math. Ser.*

PENDLEBURY (C.) and BEARD (W. S.). Arithmetic for the Standards. By C. PENDLEBURY, M.A., F.R.A.S., and W. S. BEARD, F.R.G.S. Standards I., II., III., 2d. each; IV., V., VI., 3d. each. VII., *in the Press.*

— **Elementary Arithmetic.** 3rd edition. Crown 8vo, 1s. 6d.

POPE (L. J.). Lessons in Elementary Algebra. By L. J. POPE, B.A. (Lond.), Assistant Master at the Oratory School, Birmingham. First Series, up to and including Simple Equations and Problems. Crown 8vo, 1s 6d.

PROWDE SMITH (R.). *See* Macmichael.

SHAW (S. J. D.). Arithmetic Papers. Set in the Cambridge Higher Local Examination, from June, 1869, to June, 1887, inclusive, reprinted by permission of the Syndicate By S. J D. SHAW, Mathematical Lecturer of Newnham College. Crown 8vo, 2s. 6d.; Key, 4s. 6d. net.

TAIT (T. S.). *See* Pendlebury.

WATSON (J.) and GOUDIE (W. P.). Arithmetic. A Progressive Course of Examples With Answers. By J. WATSON, M.A., Corpus Christi College, Cambridge, formerly Senior Mathematical Master of the Ordnance School, Carshalton. 7th edition, revised and enlarged. By W. P. GOUDIE, B.A. Lond. Fcap. 8vo, 2s. 6d. [*Camb. S. and C. Texts.*

WHITWORTH (W. A.). Algebra. Choice and Chance. An Elementary Treatise on Permutations, Combinations, and Probability, with 640 Exercises and Answers. By W. A. WHITWORTH, M.A., Fellow of St. John's College, Cambridge. 4th edition, revised and enlarged. Crown 8vo, 6s. [*Camb. Math. Ser.*

WRIGLEY (A.) Arithmetic. By A. WRIGLEY, M.A, St. John's College. Fcap. 8vo, 3s. 6d. [*Camb. S. and C. Texts.*

BOOK-KEEPING.

CRELLIN (P.). A New Manual of Book-keeping, combining the Theory and Practice, with Specimens of a set of Books. By PHILLIP CRELLIN, Chartered Accountant. Crown 8vo, 3s. 6d.

— **Book-keeping for Teachers and Pupils.** Crown 8vo, 1s. 6d. Key, 2s. net.

FOSTER (B. W.). Double Entry Elucidated. By B. W. FOSTER. 14th edition. Fcap. 4to, 3s. 6d.

MEDHURST (J. T.). Examination Papers in Book-keeping. Compiled by JOHN T. MEDHURST, A.K.C., F.S.S., Fellow of the Society of Accountants and Auditors, and Lecturer at the City of London College. 3rd edition. Crown 8vo, 3s.

THOMSON (A. W.). A Text-Book of the Principles and Practice of Book-keeping. By PROFESSOR A. W. THOMSON, B.SC., Royal Agricultural College, Cirencester. Crown 8vo, 5s.

GEOMETRY AND EUCLID.

BESANT (W. H.). Geometrical Conic Sections. By W. H. BESANT, SC.D., F.R.S., Fellow of St. John's College, Cambridge. *8th edition.* Fcap. 8vo, 4*s.* 6*d.* Enunciations, separately, sewed, 1*s.*
[*Camb. S. and C. Texts.*

BRASSE (J.). The Enunciations and Figures of Euclid, prepared for Students in Geometry By the REV. J. BRASSE, D.D. *New edition.* Fcap. 8vo, 1*s.* Without the Figures, 6*d.*

DEIGHTON (H.). Euclid. Books I.-VI., and part of Book XI., newly translated from the Greek Text, with Supplementary Propositions, Chapters on Modern Geometry, and numerous Exercises. By HORACE DEIGHTON, M.A., Head Master of Harrison College, Barbados. *3rd edition.* 4*s.* 6*d.* Key, for tutors only, 5*s.* net. [*Camb. Math. Ser.*
Also issued in parts :—Book I., 1*s.* ; Books I. and II., 1*s.* 6*d.* ; Books I.-III., 2*s.* 6*d.* ; Books III. and IV., 1*s.* 6*d.*

DIXON (E. T.). The Foundations of Geometry. By EDWARD T. DIXON, late Royal Artillery. Demy 8vo, 6*s.*

MASON (C. P.). Euclid. The First Two Books Explained to Beginners. By C. P. MASON, B.A. *2nd edition.* Fcap. 8vo, 2*s.* 6*d.*

McDOWELL (J.) Exercises on Euclid and in Modern Geometry, containing Applications of the Principles and Processes of Modern Pure Geometry. By the late J. MCDOWELL, M.A., F.R.A.S., Pembroke College, Cambridge, and Trinity College, Dublin. *4th edition.* 6*s.*
[*Camb. Math. Ser.*

TAYLOR (C.). An Introduction to the Ancient and Modern Geometry of Conics, with Historical Notes and Prolegomena. 15*s.*

— The Elementary Geometry of Conics. By C. TAYLOR, D.D., Master of St. John's College. *7th edition, revised.* With a Chapter on the Line Infinity, and a new treatment of the Hyperbola. Crown 8vo, 4*s.* 6*d.*
[*Camb. Math. Ser.*

WEBB (R.). The Definitions of Euclid. With Explanations and Exercises, and an Appendix of Exercises on the First Book by R. WEBB M.A. Crown 8vo, 1*s.* 6*d.*

WILLIS (H. G.). Geometrical Conic Sections. An Elementary Treatise. By H. G. WILLIS, M.A., Clare College, Cambridge, Assistant Master of Manchester Grammar School. Crown 8vo, 5*s.*
[*Camb. Math. Ser.*

ANALYTICAL GEOMETRY, ETC.

ALDIS (W. S.). Solid Geometry, An Elementary Treatise on. By W. S. ALDIS, M.A., late Professor of Mathematics in the University College, Auckland, New Zealand. *4th edition, revised.* Crown 8vo, 6*s.*
[*Camb. Math. Ser.*

BESANT (W. H.). Notes on Roulettes and Glissettes. By W. H. BESANT, SC.D., F.R.S. *2nd edition, enlarged.* Crown 8vo, 5*s.*
[*Camb. Math. Ser.*

CAYLEY (A.). Elliptic Functions, An Elementary Treatise on. By ARTHUR CAYLEY, Sadlerian Professor of Pure Mathematics in the University of Cambridge. Demy 8vo. *New edition in the Press.*

TURNBULL (W. P.). Analytical Plane Geometry. An Introduction to. By W. P. TURNBULL, M.A., sometime Fellow of Trinity College. 8vo, 12s.

VYVYAN (T. G.). Analytical Geometry for Schools. By REV. T. VYVYAN, M.A., Fellow of Gonville and Caius College, and Mathematical Master of Charterhouse. *6th edition.* 8vo, 4s. 6d.
[*Camb. S. and C. Texts.*

— Analytical Geometry for Beginners. Part I. The Straight Line and Circle. Crown 8vo, 2s. 6d. [*Camb. Math. Ser.*

WHITWORTH (W. A.). Trilinear Co-ordinates, and other methods of Modern Analytical Geometry of Two Dimensions. By W. A. WHITWORTH, M.A., late Professor of Mathematics in Queen's College, Liverpool, and Scholar of St. John's College, Cambridge. 8vo, 16s.

TRIGONOMETRY.

DYER (J. M.) and WHITCOMBE (R. H.). Elementary Trigonometry. By J. M. DYER, M.A. (Senior Mathematical Scholar at Oxford), and REV. R. H. WHITCOMBE, Assistant Masters at Eton College. *2nd edition.* Crown 8vo, 4s. 6d. [*Camb. Math. Ser.*

VYVYAN (T. G.). Introduction to Plane Trigonometry. By the REV. T. G. VYVYAN, M.A., formerly Fellow of Gonville and Caius College, Senior Mathematical Master of Charterhouse. *3rd edition, revised and augmented.* Crown 8vo, 3s. 6d. [*Camb. Math. Ser.*

WARD (G. H.). Examination Papers in Trigonometry. By G. H. WARD, M.A., Assistant Master at St. Paul's School. Crown 8vo, 2s. 6d. Key, 5s. net.

MECHANICS AND NATURAL PHILOSOPHY.

ALDIS (W. S.). Geometrical Optics, An Elementary Treatise on. By W. S. ALDIS, M.A. *4th edition.* Crown 8vo, 4s. [*Camb. Math. Ser.*

— An Introductory Treatise on Rigid Dynamics. Crown 8vo, 4s.
[*Camb. Math. Ser.*

— Fresnel's Theory of Double Refraction, A Chapter on. *2nd edition, revised.* 8vo, 2s.

BASSET (A. B.). A Treatise on Hydrodynamics, with numerous Examples. By A. B. BASSET, M.A., F.R.S., Trinity College, Cambridge. Demy 8vo. Vol. I., price 10s. 6d.; Vol. II., 12s. 6d.

— An Elementary Treatise on Hydrodynamics and Sound. Demy 8vo, 7s. 6d.

— A Treatise on Physical Optics. Demy 8vo, 16s.

BESANT (W. H.). Elementary Hydrostatics. By W. H. BESANT, SC.D., F.R.S. *16th edition.* Crown 8vo, 4s. 6d. Solutions, 5s.
[*Camb. Math. Ser.*

— Hydromechanics, A Treatise on. Part I. Hydrostatics. *5th edition revised, and enlarged.* Crown 8vo, 5s. [*Camb. Math. Ser.*

BESANT (W. H.). A Treatise on Dynamics. *2nd edition.* Crown 8vo, 10s. 6d. [*Camb. Math. Ser.*

CHALLIS (PROF.). Pure and Applied Calculation. By the late REV. J. CHALLIS, M.A., F.R.S., &c. Demy 8vo, 15s.
— Physics, The Mathematical Principle of. Demy 8vo, 5s.
— Lectures on Practical Astronomy. Demy 8vo, 10s.

EVANS (J. H.) and MAIN (P. T.). Newton's Principia, The First Three Sections of, with an Appendix; and the Ninth and Eleventh Sections. By J. H. EVANS, M.A., St. John's College. The 5th edition, edited by P. T. MAIN, M.A., Lecturer and Fellow of St. John's College. Fcap. 8vo, 4s. [*Camb. S. and C. Texts.*

GALLATLY (W.). Elementary Physics, Examples and Examination Papers in. Statics, Dynamics, Hydrostatics, Heat, Light, Chemistry, Electricity, London Matriculation, Cambridge B.A., Edinburgh, Glasgow, South Kensington, Cambridge Junior and Senior Papers, and Answers. By W. GALLATLY, M.A., Pembroke College, Cambridge, Assistant Examiner, London University. Crown 8vo, 4s. [*Camb. Math. Ser.*

GARNETT (W.). Elementary Dynamics for the use of Colleges and Schools. By WILLIAM GARNETT, M.A., D.C.L., Fellow of St. John's College, late Principal of the Durham College of Science, Newcastle-upon-Tyne. *5th edition, revised.* Crown 8vo, 6s. [*Camb. Math. Ser.*
— Heat, An Elementary Treatise on. *6th edition, revised.* Crown 8vo, 4s. 6d. [*Camb. Math. Ser.*

GOODWIN (H.). Statics. By H. GOODWIN, D.D., late Bishop of Carlisle. *2nd edition.* Fcap. 8vo, 3s. [*Camb. S. and C. Texts.*

HOROBIN (J. C.). Elementary Mechanics. Stage I. II. and III., 1s. 6d. each. By J. C. HOROBIN, M.A., Principal of Homerton New College, Cambridge.
— Theoretical Mechanics. Division I. Crown 8vo, 2s. 6d.
*** This book covers the ground of the Elementary Stage of Division I. of Subject VI. of the "Science Directory," and is intended for the examination of the Science and Art Department.

JESSOP (C. M.). The Elements of Applied Mathematics. Including Kinetics, Statics and Hydrostatics. By C. M. JESSOP, M.A., late Fellow of Clare College, Cambridge, Lecturer in Mathematics in the Durham College of Science, Newcastle-on-Tyne. Crown 8vo, 6s. [*Camb. Math. Ser.*

MAIN (P. T.). Plane Astronomy, An Introduction to. By P. T. MAIN, M.A., Lecturer and Fellow of St. John's College. *6th edition, revised.* Fcap. 8vo, 4s. [*Camb. S. and C. Texts.*

PARKINSON (R. M.). Structural Mechanics. By R. M. PARKINSON, ASSOC. M.I.C.E. Crown 8vo, 4s. 6d.

PENDLEBURY (C.). Lenses and Systems of Lenses, Treated after the Manner of Gauss. By CHARLES PENDLEBURY, M.A., F.R.A.S., Senior Mathematical Master of St. Paul's School, late Scholar of St. John's College, Cambridge. Demy 8vo, 5s.

STEELE (R. E.). Natural Science Examination Papers. By R. E. STEELE, M.A., F.C.S., Chief Natural Science Master, Bradford Grammar School. Crown 8vo. Part I., Inorganic Chemistry, 2s. 6d. Part II., Physics (Sound, Light, Heat, Magnetism, Electricity), 2s. 6d. [*School Exam. Series.*

WALTON (W.). Theoretical Mechanics, Problems in. By W. WALTON, M.A., Fellow and Assistant Tutor of Trinity Hall, Mathematical Lecturer at Magdalene College. *3rd edition, revised.* Demy 8vo, 16s.
— **Elementary Mechanics,** Problems in. *2nd edition.* Crown 8vo, 6s.
[*Camb. Math. Ser.*

DAVIS (J. F.). Army Mathematical Papers. Being Ten Years' Woolwich and Sandhurst Preliminary Papers. Edited, with Answers, by J. F. DAVIS, D.LIT., M.A. Lond. Crown 8vo, 2s. 6d.

DYER (J. M.) and PROWDE SMITH (R.). Mathematical Examples. A Collection of Examples in Arithmetic, Algebra, Trigonometry, Mensuration, Theory of Equations, Analytical Geometry, Statics, Dynamics, with Answers, &c. For Army and Indian Civil Service Candidates. By J. M. DYER, M.A., Assistant Master, Eton College (Senior Mathematical Scholar at Oxford), and R. PROWDE SMITH, M A. Crown 8vo, 6s. [*Camb. Math. Ser.*

GOODWIN (H.). Problems and Examples, adapted to "Goodwin's Elementary Course of Mathematics." By T. G. VYVYAN, M.A. *3rd edition.* 8vo, 5s.; Solutions, *3rd edition,* 8vo, 9s.

SMALLEY (G. R.). A Compendium of Facts and Formulae in Pure Mathematics and Natural Philosophy. By G. R. SMALLEY, F.R.A.S. *New edition, revised and enlarged.* By J. McDOWELL, M.A., F.R.A.S. Fcap. 8vo, 2s.

WRIGLEY (A.). Collection of Examples and Problems in Arithmetic, Algebra, Geometry, Logarithms, Trigonometry, Conic Sections, Mechanics, &c., with Answers and Occasional Hints. By the REV. A. WRIGLEY. *10th edition, 20th thousand.* Demy 8vo, 8s. 6d.
A Key. By J. C. PLATTS, M.A. and the REV. A. WRIGLEY. *2nd edition.* Demy 8vo, 10s. 6d.

MODERN LANGUAGES.

ENGLISH.

ADAMS (E.). The Elements of the English Language. By ERNEST ADAMS, PH.D. *26th edition.* Revised by J. F. DAVIS, D.LIT., M.A., (LOND.). Post 8vo, 4s. 6d.
— The Rudiments of English Grammar and Analysis. By ERNEST ADAMS, PH.D. *19th thousand.* Fcap. 8vo, 1s.

ALFORD (DEAN). The Queen's English: A Manual of Idiom and Usage. By the late HENRY ALFORD, D.D., Dean of Canterbury. *6th edition.* Small post 8vo. Sewed, 1s., cloth, 1s. 6d.

ASCHAM'S Scholemaster. Edited by PROFESSOR J. E. B. MAYOR. Small post 8vo, sewed, 1s.

BELL'S ENGLISH CLASSICS. A New Series, Edited for use in Schools, with Introduction and Notes. Crown 8vo.
  JOHNSON'S Life of Addison. Edited by F. RYLAND, Author of "The Students' Handbook of Psychology," etc. 2s. 6d.
  — Life of Swift. Edited by F. RYLAND, M.A. 2s.

BELL'S ENGLISH CLASSICS—*continued.*
JOHNSON'S Life of Pope. Edited by F. RYLAND, M.A. 2s. 6d.
— Life of Milton. Edited by F. RYLAND, M.A. 2s. 6d.
— Life of Dryden. Edited by F. RYLAND, M.A. [*Preparing.*
LAMB'S Essays. Selected and Edited by K. DEIGHTON. 3s.; sewed, 2s.
BYRON'S Childe Harold. Edited by H. G. KEENE, M.A., C.I E., Author of "A Manual of French Literature," etc. 3s. 6d. Also Cantos I. and II. separately; sewed, 1s. 9d.
— Siege of Corinth. Edited by P. HORDERN, late Director of Public Instruction in Burma. 1s. 6d; sewed, 1s.
MACAULAY'S Lays of Ancient Rome. Edited by P. HORDERN. 2s. 6d.; sewed, 1s. 9d.
MASSINGER'S A New Way to Pay Old Debts. Edited by K. DEIGHTON. 3s.; sewed, 2s.
BURKE'S Letters on a Regicide Peace. I. and II. Edited by H. G. KEENE, M.A., C.I.E. 3s.; sewed, 2s.
MILTON'S Paradise Regained. Edited by K. DEIGHTON. 2s. 6d.; sewed, 1s. 9d.
SELECTIONS FROM POPE. Containing Essay on Criticism, Rape of the Lock, Temple of Fame, Windsor Forest. Edited by K. DEIGHTON. 2s. 6d.; sewed, 1s. 9d.
GOLDSMITH'S Good-Natured Man and She Stoops to Conquer. Edited by K. DEIGHTON. Each, 2s. cloth; 1s. 6d. sewed.
DE QUINCEY, Selections from. The English Mail-Coach and The Revolt of the Tartars. Edited by CECIL M. BARROW, M.A., Principal of Victoria College, Palghât. [*In the press.*
MILTON'S Paradise Lost. Books I. and II. Edited by R. G. OXENHAM, M.A., Principal of Elphinstone College, Bombay. [*Preparing.*
— Books III. and IV. Edited by R. G. OXENHAM. [*Preparing.*
SELECTIONS FROM CHAUCER. Edited by J. B. BILDERBECK, B.A., Professor of English Literature, Presidency College, Madras. [*Preparing.*
MACAULAY'S Essay on Clive. Edited by CECIL BARROW, M.A. [*Preparing.*
BROWNING'S Strafford. Edited by E. H. HICKEY. With Introduction by S. R. GARDINER, LL.D. 2s. 6d.
SHAKESPEARE'S Julius Caesar. Edited by T. DUFF BARNETT, B.A. (Lond.). 2s.
— Merchant of Venice. Edited by T. DUFF BARNETT, B.A. (Lond.). 2s.
— Tempest. Edited by T. DUFF BARNETT, B.A. (Lond.). 2s.

Others to follow.

BELL'S READING BOOKS. Post 8vo, cloth, illustrated.

Infants.
Infant's Primer. 3d.
Tot and the Cat. 6d.
The Old Boathouse. 6d.
The Cat and the Hen. 6d.

Standard I.
School Primer. 6d.
The Two Parrots. 6d.
The Three Monkeys. 6d.
The New-born Lamb. 6d.
The Blind Boy. 6d.

Standard II.
The Lost Pigs. 6d.
Story of a Cat. 6d.
Queen Bee and Busy Bee. 6d.

Gull's Crag. 6d.
Great Deeds in English History. 1s.

Standard III.
Adventures of a Donkey. 1s.
Grimm's Tales. 1s.
Great Englishmen. 1s.
Andersen's Tales. 1s.
Life of Columbus. 1s.

Standard IV.
Uncle Tom's Cabin. 1s.
Great Englishwomen. 1s.
Great Scotsmen. 1s.
Edgeworth's Tales. 1s.
Gatty's Parables from Nature. 1s.
Scott's Talisman. 1s.

BELL'S READING BOOKS—*continued.*

Standard V.
Dickens' Oliver Twist. 1s.
Dickens' Little Nell. 1s.
Masterman Ready. 1s.
Marryat's Poor Jack. 1s.
Arabian Nights. 1s.
Gulliver's Travels. 1s.
Lyrical Poetry for Boys and Girls. 1s.
Vicar of Wakefield. 1s.

Standards VI. and VII.
Lamb's Tales from Shakespeare. 1s.
Robinson Crusoe. 1s.
Tales of the Coast. 1s.
Settlers in Canada. 1s.
Southey's Life of Nelson. 1s.
Sir Roger de Coverley. 1s.

BELL'S GEOGRAPHICAL READERS. By M. J. BARRINGTON-WARD, M A. (Worcester College, Oxford).

The Child's Geography. Illustrated. Stiff paper cover, 6d.
The Map and the Compass. (Standard I) Illustrated. Cloth, 8d.

The Round World. (Standard II.) Illustrated. Cloth, 10d.
About England. (Standard III.) With Illustrations and Coloured Map. Cloth, 1s. 4d.

EDWARDS (F.). Examples for Analysis in Verse and Prose from well-known sources, selected and arranged by F. EDWARDS. *New edition.* Fcap. 8vo, cloth, 1s.

GOLDSMITH. The Deserted Village. Edited, with Notes and Life, by C. P. MASON, B.A., F.C.P. *4th edition.* Crown 8vo, 1s.

HANDBOOKS OF ENGLISH LITERATURE. Edited by J. W. HALES, M.A., formerly Clark Lecturer in English Literature at Trinity College, Cambridge, Professor of English Literature at King's College, London. Crown 8vo, 3s. 6d. each.

The Age of Pope. By JOHN DENNIS. [*Ready*.

In preparation.

The Age of Chaucer. By PROFESSOR HALES.
The Age of Shakespeare. By PROFESSOR HALES.
The Age of Milton. By J. BASS MULLINGER, M.A.
The Age of Dryden. By W. GARNETT, LL.D.
The Age of Wordsworth. By PROFESSOR C. H. HERFORD, LITT.D.

Other volumes to follow.

HAZLITT (W.). Lectures on the Literature of the Age of Elizabeth. Small post 8vo, sewed, 1s.

— Lectures on the English Poets. Small post 8vo, sewed, 1s.

— Lectures on the English Comic Writers. Small post 8vo, sewed, 1s.

LAMB (C.). Specimens of English Dramatic Poets of the Time of Elizabeth. With Notes, together with the Extracts from the Garrick Plays.

MASON (C. P.). Grammars by C. P. MASON, B.A., F.C.P., Fellow of University College, London.

— First Notions of Grammar for Young Learners. Fcap. 8vo. 85*th thousand.* Cloth, 1s.

— First Steps in English Grammar, for Junior Classes. Demy 18mo. 54*th thousand.* 1s.

MASON (C. P.). Outlines of English Grammar, for the Use of Junior Classes. *17th edition*. *97th thousand*. Crown 8vo, 2s.
— English Grammar; including the principles of Grammatical Analysis. *35th edition, revised*. *148th thousand*. Crown 8vo, green cloth, 3s. 6d.
— A Shorter English Grammar, with copious and carefully graduated Exercises, based upon the author's English Grammar. *9th edition*. *49th thousand*. Crown 8vo, brown cloth, 3s. 6d.
— Practice and Help in the Analysis of Sentences. Price 2s. Cloth.
— English Grammar Practice, consisting of the Exercises of the Shorter English Grammar published in a separate form. *3rd edition*. Crown 8vo, 1s.
— Remarks on the Subjunctive and the so-called Potential Mood. 6d., sewn.
— Blank Sheets Ruled and headed for Analysis. 1s. per dozen.
MILTON: Paradise Lost. Books I., II., and III. Edited, with Notes on the Analysis and Parsing, and Explanatory Remarks, by C. P. MASON, B.A., F.C.P. Crown 8vo.
    Book I. With Life. *5th edition*. 1s.
    Book II. With Life. *3rd edition*. 1s.
    Book III. With Life. *2nd edition*. 1s.
— Paradise Lost. Books V-VIII. With Notes for the Use of Schools. By C. M. LUMBY. 2s. 6d..
PRICE (A. C.). Elements of Comparative Grammar and Philology. For Use in Schools. By A. C. PRICE, M.A., Assistant Master at Leeds Grammar School; late Scholar of Pembroke College, Oxford. Crown 8vo, 2s. 6d.
SHAKESPEARE. Notes on Shakespeare's Plays. With Introduction, Summary, Notes (Etymological and Explanatory), Prosody, Grammatical Peculiarities, etc. By T. DUFF BARNETT, B.A. Lond., late Second Master in the Brighton Grammar School. Specially adapted for the Local and Preliminary Examinations. Crown 8vo, 1s. each.
    Midsummer Night's Dream.—Julius Cæsar.—The Tempest.—Macbeth.—Henry V.—Hamlet.—Merchant of Venice.—King Richard II.—King John.—King Lear.—Coriolanus.
    "The Notes are comprehensive and concise."—*Educational Times*.
    "Comprehensive, practical, and reliable."—*Schoolmaster*.
— Hints for Shakespeare-Study. Exemplified in an Analytical Study of Julius Cæsar. By MARY GRAFTON MOBERLY. *2nd edition*. Crown 8vo, sewed, 1s.
— Coleridge's Lectures and Notes on Shakespeare and other English Poets. Edited by T. ASHE, B.A. Small post 8vo, 3s. 6d.
— Shakespeare's Dramatic Art. The History and Character of Shakespeare's Plays. By DR. HERMANN ULRICI. Translated by L. DORA SCHMITZ. 2 vols. small post 8vo, 3s. 6d. each.
— William Shakespeare. A Literary Biography. By KARL ELZE, PH.D., LL.D. Translated by L. DORA SCHMITZ. Small post 8vo, 5s.
— Hazlitt's Lectures on the Characters of Shakespeare's Plays. Small post 8vo, 1s.
See BELL'S ENGLISH CLASSICS.

SKEAT (W. W.). Questions for Examinations in English Literature. With a Preface containing brief hints on the study of English. Arranged by the REV. W. W. SKEAT, LITT. D., Elrington and Bosworth Professor of Anglo-Saxon in the University of Cambridge. *3rd edition.* Crown 8vo, 2s. 6d.

SMITH (C. J.) Synonyms and Antonyms of the English Language. Collected and Contrasted by the VEN. C. J. SMITH, M.A. *2nd edition, revised.* Small post 8vo, 5s.

— Synonyms Discriminated. A Dictionary of Synonymous Words in the English Language. Illustrated with Quotations from Standard Writers. By the late VEN. C. J. SMITH, M.A. With the Author's latest Corrections and Additions, edited by the REV. H. PERCY SMITH, M.A., of Balliol College, Oxford, Vicar of Great Barton, Suffolk. *4th edition.* Demy 8vo, 14s.

TEN BRINK'S History of English Literature. Vol. I. Early English Literature (to Wiclif). Translated into English by HORACE M. KENNEDY, Professor of German Literature in the Brooklyn Collegiate Institute. Small post 8vo, 3s. 6d.

— Vol. II. (Wiclif, Chaucer, Earliest Drama, Renaissance). Translated by W. CLARKE ROBINSON, PH.D. Small post 8vo, 3s. 6d.

THOMSON: Spring. Edited by C. P. MASON, B.A., F.C.P. With Life. *2nd edition.* Crown 8vo, 1s.

— Winter. Edited by C. P. MASON, B.A., F.C.P. With Life. Crown 8vo, 1s.

WEBSTER'S INTERNATIONAL DICTIONARY of the English Language. Including Scientific, Technical, and Biblical Words and Terms, with their Significations, Pronunciations, Alternative Spellings, Derivations, Synonyms, and numerous illustrative Quotations, with various valuable literary Appendices, with 83 extra pages of Illustrations grouped and classified, rendering the work a COMPLETE LITERARY AND SCIENTIFIC REFERENCE-BOOK. *New edition* (1890). Thoroughly revised and enlarged under the supervision of NOAH PORTER, D.D., LL.D. 1 vol. (2,118 pages, 3,500 woodcuts), 4to, cloth, 31s. 6d.; half calf, £2 2s.; half russia, £2 5s.; calf, £2 8s.; or in 2 vols. cloth, £1 14s.

Prospectuses, with specimen pages, sent post free on application.

WEBSTERS BRIEF INTERNATIONAL DICTIONARY. A Pronouncing Dictionary of the English Language, abridged from Webster's International Dictionary. With a Treatise on Pronunciation, List of Prefixes and Suffixes, Rules for Spelling, a Pronouncing Vocabulary of Proper Names in History, Geography, and Mythology, and Tables of English and Indian Money, Weights, and Measures. With 564 pages and 800 Illustrations. Demy 8vo, 3s.

WRIGHT (T.). Dictionary of Obsolete and Provincial English. Containing Words from the English Writers previous to the 19th century, which are no longer in use, or are not used in the same sense, and Words which are now used only in the Provincial Dialects. Compiled by THOMAS WRIGHT, M.A., F.S.A., etc. 2 vols. 5s. each.

FRENCH CLASS BOOKS.

BOWER (A. M.). The Public Examination French Reader. With a Vocabulary to every extract, suitable for all Students who are preparing for a French Examination. By A. M. BOWER, F.R.G.S., late Master in University College School, etc. Cloth, 3s. 6d.

BARBIER (PAUL). A Graduated French Examination Course. By PAUL BARBIER, Lecturer in the South Wales University College, etc. Crown 8vo, 3s.

BARRERE (A.) Junior Graduated French Course. Affording Materials for Translation, Grammar, and Conversation. By A. BARRÈRE, Professor R.M.A., Woolwich. 1s. 6d.
— Elements of French Grammar and First Steps in Idioms. With numerous Exercises and a Vocabulary. Being an Introduction to the Précis of Comparative French Grammar. Crown 8vo, 2s.
— Précis of Comparative French Grammar and Idioms and Guide to Examinations. 4th edition. 3s. 6d.
— Récits Militaires. From Valmy (1792) to the Siege of Paris (1870). With English Notes and Biographical Notices. 2nd edition. Crown 8vo, 3s.

CLAPIN (A. C.). French Grammar for Public Schools. By the REV. A. C. CLAPIN, M.A., St. John's College, Cambridge, and Bachelier-ès-lettres of the University of France. Fcap. 8vo. 13th edition. 2s. 6d. Key to the Exercises. 3s. 6d. net.
— French Primer. Elementary French Grammar and Exercises for Junior Forms in Public and Preparatory Schools. Fcap. 8vo. 10th edition. 1s.
— Primer of French Philology. With Exercises for Public Schools. 6th edition. Fcap. 8vo, 1s.
— English Passages for Translation into French. Crown 8vo, 2s. 6d. Key (for Tutors only), 4s. net.

DAVIS (J. F.) Army Examination Papers in French. Questions set at the Preliminary Examinations for Sandhurst and Woolwich, from Nov., 1876, to June, 1890, with Vocabulary. By J. F. DAVIS, D.LIT., M.A., Lond. Crown 8vo, 2s. 6d.

DAVIS (J. F.) and THOMAS (F.). An Elementary French Reader. Compiled, with a Vocabulary, by J. F. DAVIS, M.A., D.LIT., and FERDINAND THOMAS, Assistant Examiners in the University of London. Crown 8vo, 2s.

DELILLE'S GRADUATED FRENCH COURSE.
The Beginner's own French Book. 2s. Key, 2s.
Easy French Poetry for Beginners. 2s.
French Grammar. 3s. Key, 3s.
Repertoire des Prosateurs. 3s. 6d.
Modèles de Poesie. 3s. 6d.
Manuel Etymologique. 2s. 6d.
Synoptical Table of French Verbs. 6d.

GASC (F. E. A.). First French Book; being a New, Practical, and Easy Method of Learning the Elements of the French Language. Reset and thoroughly revised. 116th thousand. Crown 8vo, 1s.
— Second French Book; being a Grammar and Exercise Book, on a new and practical plan, and intended as a sequel to the "First French Book." 52nd thousand. Fcap. 8vo, 1s. 6d.

GASC (F. E. A.). Key to First and Second French Books. *5th edition*, Fcap. 8vo, 3*s*. 6*d*. net.
— **French Fables**, for Beginners, in Prose, with an Index of all the Words at the end of the work. *16th thousand*. 12mo, 1*s*. 6*d*.
— **Select Fables of La Fontaine.** *19th thousand.* Fcap. 8vo, 1*s*. 6*d*.
— **Histoires Amusantes et Instructives**; or, Selections of Complete Stories from the best French modern authors, who have written for the young. With English notes. *17th thousand.* Fcap. 8vo, 2*s*.
— **Practical Guide to Modern French Conversation**, containing:— I. The most current and useful Phrases in Everyday Talk. II. Everybody's necessary Questions and Answers in Travel-Talk. *19th edition.* Fcap. 8vo, 1*s*. 6*d*.
— **French Poetry for the Young.** With Notes, and preceded by a few plain Rules of French Prosody. *5th edition, revised.* Fcap. 8vo, 1*s*. 6*d*.
— **French Prose Composition**, Materials for. With copious footnotes, and hints for idiomatic renderings. *21st thousand.* Fcap. 8vo, 3*s*.
    Key. *2nd edition.* 6*s*. net.
— **Prosateurs Contemporains**; or, Selections in Prose chiefly from contemporary French literature. With notes. *11th edition.* 12mo, 3*s*. 6*d*.
— **Le Petit Compagnon**; a French Talk-Book for Little Children. *14th edition.* 16mo, 1*s*. 6*d*.
— **French and English Dictionary**, with upwards of Fifteen Thousand new words, senses, &c., hitherto unpublished. *5th edition, with numerous additions and corrections.* In one vol. 8vo, cloth, 10*s*. 6*d*. In use at Harrow, Rugby, Shrewsbury, &c.
— **Pocket Dictionary** of the French and English Languages; for the everyday purposes of Travellers and Students. Containing more than Five Thousand modern and current words, senses, and idiomatic phrases and renderings, not found in any other dictionary of the two languages. *New edition. 51st thousand.* 16mo, cloth, 2*s*. 6*d*.

GOSSET (A.). Manual of French Prosody for the use of English Students. By ARTHUR GOSSET, M.A., Fellow of New College, Oxford. Crown 8vo, 3*s*.

"This is the very book we have been looking for. We hailed the title with delight, and were not disappointed by the perusal. The reader who has mastered the contents will know, what not one in a thousand of Englishmen who read French knows, the rules of French poetry."— *Journal of Education.*

LE NOUVEAU TRESOR; designed to facilitate the Translation of English into French at Sight. By M. E. S. *18th edition.* Fcap. 8vo, 1*s*. 6*d*.

STEDMAN (A. M. M.). French Examination Papers in Miscellaneous Grammar and Idioms. Compiled by A. M. M. STEDMAN, M A. *5th edition.* Crown 8vo, 2*s*. 6*d*.
    A Key. By G. A. SCHRUMPF. For Tutors only. 6*s*. net.
    Easy French Passages for Unseen Translation. Fcap. 8vo, 1*s*. 6*d*.
— Easy French Exercises on Elementary Syntax. Crown 8vo, 2*s*. 6*d*.
— First French Lessons. Crown 8vo, 1*s*.
— French Vocabularies for Repetition. Fcap. 8vo, 1*s*.
    Steps to French. 12mo, 8*d*.

FRENCH ANNOTATED EDITIONS.

BALZAC. Ursule Mirouët. By HONORÉ DE BALZAC. Edited, with Introduction and Notes, by JAMES BOIELLE, B.-ès-L., Senior French Master, Dulwich College. 3*s*.

CLARÉTIE. Pierrille. By JULES CLARÉTIE. With 27 Illustrations. Edited, with Introduction and Notes, by JAMES BOÏELLE, B.-ès-L. 2*s*. 6*d*.

DAUDET. La Belle Nivernaise. Histoire d'un vieux bateau et de son équipage. By ALPHONSE DAUDET. Edited, with Introduction and Notes, by JAMES BOÏELLE, B.-ès-L. With Six Illustrations. 2*s*.

FÉNELON. Aventures de Télémaque. Edited by C. J. DELILLE. 4*th edition*. Fcap. 8vo, 2*s*. 6*d*.

GOMBERT'S FRENCH DRAMA. Re-edited, with Notes, by F. E. A. GASC. Sewed, 6*d*. each.

MOLIÈRE.

Le Misanthrope.
L'Avare.
Le Bourgeois Gentilhomme.
Le Tartuffe.
Le Malade Imaginaire.
Les Femmes Savantes.

Les Fourberies de Scapin.
Les Précieuses Ridicules.
L'Ecole des Femmes.
L'Ecole des Maris.
Le Médecin Malgré Lui.

RACINE.

La Thébaïde, ou Les Frères Ennemis.
Andromaque.
Les Plaideurs.
Iphigénie.

Britannicus.
Phèdre.
Esther.
Athalie.

CORNEILLE.

Le Cid.
Horace.

Cinna.
Polyeucte.

VOLTAIRE.—Zaïre.

GREVILLE. Le Moulin Frappier. By HENRY GREVILLE. Edited, with Introduction and Notes, by JAMES BOIELLE, B.-ès-L. 3*s*.

HUGO. Bug Jargal. Edited, with Introduction and Notes, by JAMES BOÏELLE, B.-ès-L. 3*s*.

LA FONTAINE. Select Fables. Edited by F. E. A. GASC. 19*th thousand*. Fcap. 8vo, 1*s*. 6*d*.

LAMARTINE. Le Tailleur de Pierres de Saint-Point. Edited with Notes by JAMES BOIELLE, B.-ès-L. 6*th thousand*. Fcap. 8vo, 1*s*. 6*d*.

SAINTINE. Picciola. Edited by DR. DUBUC. 16*th thousand*. Fcap. 8vo, 1*s*. 6*d*.

VOLTAIRE. Charles XII. Edited by L. DIREY. 7*th edition*. Fcap. 8vo, 1*s*. 6*d*.

GERMAN CLASS BOOKS.

BUCHHEIM (DR. C. A.). German Prose Composition. Consisting of Selections from Modern English Writers. With grammatical notes, idiomatic renderings, and general introduction. By C. A. BUCHHEIM, PH.D., Professor of the German Language and Literature in King's College, and

Examiner in German to the London University. 14*th edition, enlarged and revised.* With a list of subjects for original composition. Fcap. 8vo, 4*s.* 6*d.* A KEY to the 1st and 2nd parts. 3*rd edition.* 3*s.* net. To the 3rd and 4th parts. 4*s.* net.

BUCHHEIM (DR. C. A.). First Book of German Prose. Being Parts I. and II. of the above. With Vocabulary by H. R. Fcap. 8vo, 1*s.* 6*d.*

CLAPIN (A. C.). A German Grammar for Public Schools. By the REV. A. C. CLAPIN, and F. HOLL-MÜLLER, Assistant Master at the Bruton Grammar School. 6*th edition.* Fcap. 8vo, 2*s.* 6*d.*

—— A German Primer. With Exercises. 2*nd edition.* Fcap. 8vo, 1*s.*

German. The Candidate's Vade Mecum. Five Hundred Easy Sentences and Idioms. By an Army Tutor. Cloth, 1*s.* For Army Prelim. Exam.

LANGE (F.). A Complete German Course for Use in Public Schools. By F. LANGE, PH.D., Professor R.M.A. Woolwich, Examiner in German to the College of Preceptors, London ; Examiner in German at the Victoria University, Manchester. Crown 8vo.

 Concise German Grammar. With special reference to Phonology, Comparative Philology, English and German Equivalents and Idioms. Comprising Materials for Translation, Grammar, and Conversation. Elementary, 2*s.* ; Intermediate, 2*s.* ; Advanced, 3*s.* 6*d.*

 Progressive German Examination Course. Comprising the Elements of German Grammar, an Historic Sketch of the Teutonic Languages, English and German Equivalents, Materials for Translation, Dictation, Extempore Conversation, and Complete Vocabularies. I. Elementary Course, 2*s.* II. Intermediate Course, 2*s.* III. Advanced Course. *Second revised edition.* 1*s.* 6*d.*

 Elementary German Reader. A Graduated Collection of Readings in Prose and Poetry. With English Notes and a Vocabulary. 4*th edition.* 1*s.* 6*d.*

 Advanced German Reader. A Graduated Collection of Readings in Prose and Poetry. With English Notes by F. LANGE, PH.D., and J. F. DAVIS, D.LIT. 2*nd edition.* 3*s.*

MORICH (R. J.). German Examination Papers in Miscellaneous Grammar and Idioms. By R. J. MORICH, Manchester Grammar School. 2*nd edition.* Crown 8vo, 2*s.* 6*d.* A Key, for Tutors only. 5*s.* net.

STOCK (DR.). Wortfolge, or Rules and Exercises on the order of Words in German Sentences. With a Vocabulary. By the late FREDERICK STOCK, D.LIT., M.A. Fcap. 8vo, 1*s.* 6*d.*

KLUGE'S Etymological Dictionary of the German Language. Translated by J. F. DAVIS, D.LIT. (Lond.). Crown 4to, 18*s.*

GERMAN ANNOTATED EDITIONS.

AUERBACH (B.). Auf Wache. Novelle von BERTHOLD AUERBACH. Der Gefrorene Kuss. Novelle von OTTO ROQUETTE. Edited by A. A. MACDONELL, M.A., PH.D. 2*nd edition.* Crown 8vo, 2*s.*

BENEDIX (J. R.). Doktor Wespe. Lustspiel in fünf Aufzügen von JULIUS RODERICH BENEDIX. Edited by PROFESSOR F. LANGE, PH.D. Crown 8vo, 2*s.* 6*d.*

EBERS (G.). Eine Frage. Idyll von GEORG EBERS. Edited by F. STORR, B.A., Chief Master of Modern Subjects in Merchant Taylors' School. Crown 8vo, 2s.

FREYTAG (G.). Die Journalisten. Lustspiel von GUSTAV FREYTAG. Edited by PROFESSOR F. LANGE, PH.D. *4th revised edition.* Crown 8vo, 2s. 6d.

— SOLL UND HABEN. Roman von GUSTAV FREYTAG. Edited by W. HANBY CRUMP, M.A. Crown 8vo, 2s. 6d.

GERMAN BALLADS from Uhland, Goethe, and Schiller. With Introductions, Copious and Biographical Notices. Edited by C. L. BIELEFELD. *4th edition.* Fcap. 8vo, 1s. 6d.

GERMAN EPIC TALES IN PROSE. I. Die Nibelungen, von A. F. C. VILMAR. II. Walther und Hildegund, von ALBERT RICHTER. Edited by KARL NEUHAUS, PH.D., the International College, Isleworth. Crown 8vo, 2s. 6d.

GOETHE. Hermann und Dorothea. With Introduction, Notes, and Arguments. By E. BELL, M.A., and E. WÖLFEL. *2nd edition.* Fcap. 8vo, 1s. 6d.

GOETHE FAUST. Part I. German Text with Hayward's Prose Translation and Notes. Revised, With Introduction by C. A. BUCHHEIM, PH.D., Professor of German Language and Literature at King's College, London. Small post 8vo, 5s.

GUTZKOW (K.). Zopf und Schwert. Lustspiel von KARL GUTZKOW. Edited by PROFESSOR F. LANGE, PH.D. Crown 8vo, 2s. 6d.

HEY'S FABELN FÜR KINDER. Illustrated by O. SPECKTER. Edited, with an Introduction, Grammatical Summary, Words, and a complete Vocabulary, by PROFESSOR F. LANGE, PH.D. Crown 8vo, 1s. 6d.

— The same. With a Phonetic Introduction, and Phonetic Transcription of the Text. By PROFESSOR F. LANGE, PH.D. Crown 8vo, 2s.

HEYSE (P.). Hans Lange. Schauspiel von PAUL HEYSE. Edited by A. A. MACDONELL, M.A., PH.D., Taylorian Teacher, Oxford University. Crown 8vo, 2s.

HOFFMANN (E. T. A.). Meister Martin, der Küfner. Erzählung von E. T. A. HOFFMANN. Edited by F. LANGE, PH.D. *2nd edition.* Crown 8vo, 1s. 6d.

MOSER (G. VON). Der Bibliothekar. Lustspiel von G. VON MOSER. Edited by F. LANGE, PH.D. *4th edition.* Crown 8vo, 2s.

ROQUETTE (O.). *See* Auerbach.

SCHEFFEL (V. VON). Ekkehard. Erzählung des zehnten Jahrhunderts, von VICTOR VON SCHEFFEL. Abridged edition, with Introduction and Notes by HERMAN HAGER, PH.D., Lecturer in the German Language and Literature in The Owens College, Victoria University, Manchester. Crown 8vo, 3s.

SCHILLER'S Wallenstein. Complete Text, comprising the Weimar Prologue, Lager, Piccolomini, and Wallenstein's Tod. Edited by DR. BUCHHEIM, Professor of German in King's College, London. *6th edition.* Fcap. 8vo, 5s. Or the Lager and Piccolomini, 2s. 6d. Wallenstein's Tod, 2s. 6d.

— **Maid of Orleans.** With English Notes by DR. WILHELM WAGNER. *3rd edition.* Fcap. 8vo, 1s. 6d.

— **Maria Stuart.** Edited by V. KASTNER, B.-ès-L., Lecturer on French Language and Literature at Victoria University, Manchester. *3rd edition.* Fcap. 8vo, 1s. 6d.

c

ITALIAN.

CLAPIN (A. C.). Italian Primer. With Exercises. By the REV. A. C. CLAPIN, M.A., B.-ès-L. *3rd edition.* Fcap. 8vo, 1s.

DANTE. The Inferno. A Literal Prose Translation, with the Text of the Original collated with the best editions, printed on the same page, and Explanatory Notes. By JOHN A. CARLYLE, M.D. With Portrait. *2nd edition.* Small post 8vo, 5s.

— The Purgatorio. A Literal Prose Translation, with the Text of Bianchi printed on the same page, and Explanatory Notes. By W. S. DUGDALE. Small post 8vo, 5s.

BELL'S MODERN TRANSLATIONS.

A Series of Translations from Modern Languages, with Memoirs, Introductions, etc. Crown 8vo, 1s. each.

GOETHE. Egmont. Translated by ANNA SWANWICK. With Memoir.
— Iphigenia in Tauris. Translated by ANNA SWANWICK. With Memoir.
HAUFF. The Caravan. Translated by S. MENDEL. With Memoir.
— The Inn in the Spessart. Translated by S. MENDEL. With Memoir.
LESSING. Laokoon. Translated by E. C. BEASLEY. With Memoir.
— Nathan the Wise. Translated by R. DILLON BOYLAN. With Memoir.
— Minna von Barnhelm. Translated by ERNEST BELL, M.A. With Memoir.
MOLIÈRE. The Misanthrope. Translated by C. HERON WALL. With Memoir.
— The Doctor in Spite of Himself. (Le Médecin malgré lui). Translated by C. HERON WALL. With Memoir.
— Tartuffe; or, The Impostor. Translated by C. HERON WALL. With Memoir.
— The Miser. (L'Avare). Translated by C. HERON WALL. With Memoir.
— The Shopkeeper turned Gentleman. (Le Bourgeois Gentilhomme). Translated by C. HERON WALL. With Memoir.
RACINE. Athalie. Translated by R. BRUCE BOSWELL, M.A. With Memoir.
— Esther. Translated by R. BRUCE BOSWELL, M.A. With Memoir.
SCHILLER. William Tell. Translated by SIR THEODORE MARTIN, K.C.B., LL.D. *New edition, entirely revised.* With Memoir.
— The Maid of Orleans. Translated by ANNA SWANWICK. With Memoir.
— Mary Stuart. Translated by J. MELLISH. With Memoir.

⁂ For other Translations of Modern Languages, see the Catalogue of Bohn's Libraries, which will be forwarded on application.

SCIENCE, TECHNOLOGY, AND ART.
CHEMISTRY.

STÖCKHARDT (J. A.). Experimental Chemistry. Founded on the work of J. A. STÖCKHARDT. A Handbook for the Study of Science by Simple Experiments. By C. W. HEATON, F.I.C., F.C.S., Lecturer in Chemistry in the Medical School of Charing Cross Hospital, Examiner in Chemistry to the Royal College of Physicians, etc. *Revised edition.* 5s.

Educational Catalogue. 35

WILLIAMS (W. M.). **The Framework of Chemistry.** Part I. Typical Facts and Elementary Theory. By W. M. WILLIAMS, M.A., St. John's College, Oxford; Science Master, King Henry VIII.'s School, Coventry. Crown 8vo, paper boards, 9*d.* net.

BOTANY.

EGERTON-WARBURTON (G.). **Names and Synonyms of British Plants.** By the REV. G. EGERTON-WARBURTON. Fcap. 8vo, 3*s.* 6*d.* (*Uniform with Hayward's Botanist's Pocket Book.*)

HAYWARD (W. R.). **The Botanist's Pocket-Book.** Containing in a tabulated form, the chief characteristics of British Plants, with the botanical names, soil, or situation, colour, growth, and time of flowering of every plant, arranged under its own order; with a copious Index. By W. R. HAYWARD. *6th edition, revised.* Fcap. 8vo, cloth limp, 4*s.* 6*d.*

MASSEE (G.). **British Fungus-Flora.** A Classified Text-Book of Mycology. By GEORGE MASSEE, Author of "The Plant World." With numerous Illustrations. 3 vols. post 8vo. Vols. I., II., and III. ready, 7*s.* 6*d.* each. Vol. IV. in the Press.

SOWERBY'S **English Botany.** Containing a Description and Life-size Drawing of every British Plant. Edited and brought up to the present standard of scientific knowledge, by T. BOSWELL (late SYME), LL.D., F.L.S., etc. *3rd edition, entirely revised.* With Descriptions of all the Species by the Editor, assisted by N. E. BROWN. 12 vols., with 1,937 *coloured plates,* £24 3*s.* in cloth, £26 11*s.* in half-morocco, and £30 9*s.* in whole morocco. Also in 89 parts, 5*s.*, except Part 89, containing an Index to the whole work, 7*s.* 6*d.*

*** A Supplement, to be completed in 8 or 9 parts, is now publishing. Parts I., II., and III. ready, 5*s.* each, or bound together, making Vol. XIII. of the complete work, 17*s.*

TURNBULL (R.). **Index of British Plants**, according to the London Catalogue (Eighth Edition), including the Synonyms used by the principal authors, an Alphabetical List of English Names, etc. By ROBERT TURNBULL. Paper cover, 2*s.* 6*d.*, cloth, 3*s.*

GEOLOGY.

JUKES-BROWNE (A. J.). **Student's Handbook of Physical Geology.** By A. J. JUKES-BROWNE, B.A., F.G.S., of the Geological Survey of England and Wales. With numerous Diagrams and Illustrations. *2nd edition, much enlarged,* 7*s.* 6*d.*

— **Student's Handbook of Historical Geology.** With numerous Diagrams and Illustrations. 6*s.*

"An admirably planned and well executed 'Handbook of Historical Geology.'"—*Journal of Education.*

— **The Building of the British Isles.** A Study in Geographical Evolution. With Maps. *2nd edition revised.* 7*s.* 6*d.*

MEDICINE.

CARRINGTON (R. E.), and LANE (W. A.). A Manual of Dissections of the Human Body. By the late R. E. CARRINGTON, M.D. (Lond.), F.R.C.P., Senior Assistant Physician, Guy's Hospital. 2nd edition. Revised and enlarged by W. ARBUTHNOT LANE, M.S., F.R.C.S., Assistant Surgeon to Guy's Hospital, etc. Crown 8vo, 9s.

" As solid a piece of work as ever was put into a book; accurate from beginning to end, and unique of its kind."—*British Medical Journal.*

HILTON'S Rest and Pain. Lectures on the Influence of Mechanical and Physiological Rest in the Treatment of Accidents and Surgical Diseases, and the Diagnostic Value of Pain. By the late JOHN HILTON, F.R.S., F.R.C.S., etc. Edited by W. H. A. JACOBSON, M.A., M.CH. (Oxon.), F.R.C.S. 5th edition. 9s.

HOBLYN'S Dictionary of Terms used in Medicine and the Collateral Sciences. 12th edition. Revised and enlarged by J. A. P. PRICE, B.A., M.D. (Oxon.). 10s. 6d.

LANE (W. A.). Manual of Operative Surgery. For Practitioners and Students. By W. ARBUTHNOT LANE, M.B., M.S., F.R.C.S., Assistant Surgeon to Guy's Hospital. Crown 8vo, 8s. 6d.

SHARP (W.) Therapeutics founded on Antipraxy. By WILLIAM SHARP, M.D., F.R.S. Demy 8vo, 6s.

BELL'S AGRICULTURAL SERIES.

In crown 8vo, Illustrated, 160 pages, cloth, 2s. 6d. each.

CHEAL (J.). Fruit Culture. A Treatise on Planting, Growing, Storage of Hardy Fruits for Market and Private Growers. By J. CHEAL, F.R.H.S., Member of Fruit Committee, Royal Hort. Society, etc.

FREAM (DR.). Soils and their Properties. By DR. WILLIAM FREAM, B.SC. (Lond.), F.L.S., F.G.S., F.S.S., Associate of the Surveyor's Institution, Consulting Botanist to the British Dairy Farmers' Association and the Royal Counties Agricultural Society; Prof. of Nat. Hist. in Downton College, and formerly in the Royal Agric. Coll., Cirencester.

GRIFFITHS (DR.). Manures and their Uses. By DR. A. B. GRIFFITHS, F.R.S.E., F.C.S., late Principal of the School of Science, Lincoln; Membre de la Société Chimique de Paris; Author of "A Treatise on Manures," etc., etc. *In use at Downton College.*

— **The Diseases of Crops and their Remedies.**

MALDEN (W. J.). Tillage and Implements. By W. J. MALDEN, Prof. of Agriculture in the College, Downton.

SHELDON (PROF.). The Farm and the Dairy. By PROFESSOR J. P. SHELDON, formerly of the Royal Agricultural College, and of the Downton College of Agriculture, late Special Commissioner of the Canadian Government. *In use at Downton College.*

Specially adapted for Agricultural Classes. Crown 8vo. Illustrated. 1s. each.

Practical Dairy Farming. By PROFESSOR SHELDON. Reprinted from the Author's larger work entitled "The Farm and the Dairy."

Practical Fruit Growing. By J. CHEAL, F.R.H.S. Reprinted from the author's larger work, entitled "Fruit Culture."

Educational Catalogue. 37

TECHNOLOGICAL HANDBOOKS.
Edited by Sir H. Trueman Wood.
Specially adapted for candidates in the examinations of the City Guilds Institute. Illustrated and uniformly printed in small post 8vo.

BEAUMONT (R.). Woollen and Worsted Cloth Manufacture. By ROBERTS BEAUMONT, Professor of Textile Industry, Yorkshire College, Leeds; Examiner in Cloth Weaving to the City and Guilds of London Institute. 2nd edition. 7s. 6d.

BENEDIKT (R), and KNECHT (E.). Coal-tar Colours, The Chemistry of. With special reference to their application to Dyeing, etc. By DR. R. BENEDIKT, Professor of Chemistry in the University of Vienna. Translated by E. KNECHT, PH.D. of the Technical College, Bradford. *2nd and enlarged edition*, 6s. 6d.

CROOKES (W.). Dyeing and Tissue-Printing. By WILLIAM CROOKES, F.R.S., V.P.C.S. 5s.

GADD (W. L.). Soap Manufacture. By W. LAWRENCE GADD, F.I.C., F.C.S., Registered Lecturer on Soap-Making and the Technology of Oils and Fats, also on Bleaching, Dyeing, and Calico Printing, to the City and Guilds of London Institute. 5s.

HELLYER (S. S.). Plumbing: Its Principles and Practice. By S. STEVENS HELLYER. With numerous Illustrations. 5s.

HORNBY (J.). Gas Manufacture. By J. HORNBY, F.I.C., Lecturer under the City and Guilds of London Institute. *[Preparing.*

HURST (G.H.). Silk-Dyeing and Finishing. By G. H. HURST, F.C.S., Lecturer at the Manchester Technical School, Silver Medallist, City and Guilds of London Institute. With Illustrations and numerous Coloured Patterns. 7s. 6d.

JACOBI (C. T.). Printing. A Practical Treatise. By C. T. JACOBI, Manager of the Chiswick Press, Examiner in Typography to the City and Guilds of London Institute. With numerous Illustrations. 5s.

MARSDEN (R.). Cotton Spinning: Its Development, Principles, and Practice, with Appendix on Steam Boilers and Engines. By R. MARSDEN, Editor of the "Textile Manufacturer." *4th edition*. 6s. 6d.
— **Cotton Weaving** With numerous Illustrations. *[In the press.*

POWELL (H.), CHANCE (H.), and HARRIS (H. G.). Glass Manufacture. Introductory Essay, by H. POWELL, B.A. (Whitefriars Glass Works); **Sheet Glass,** by HENRY CHANCE, M.A. (Chance Bros., Birmingham); **Plate Glass,** by H. G. HARRIS, Assoc. Memb. Inst. C.E. 3s. 6d.

ZAEHNSDORF (J. W.) Bookbinding. By J. W. ZAEHNSDORF, Examiner in Bookbinding to the City and Guilds of London Institute. With 8 Coloured Plates and numerous Diagrams. *2nd edition, revised and enlarged*. 5s.

** *Complete List of Technical Books on Application.*

MUSIC.
BANISTER (H. C.). A Text Book of Music: By H. C. BANISTER, Professor of Harmony and Composition at the R.A. of Music, at the Guild-

BANISTER (H. C.)—*continued.*
hall School of Music, and at the Royal Normal Coll. and Acad. of Music for the Blind. 15*th edition.* Fcap. 8vo. 5*s.*
This Manual contains chapters on Notation, Harmony, and Counterpoint; Modulation, Rhythm, Canon, Fugue, Voices, and Instruments ; together with exercises on Harmony, an Appendix of Examination Papers, and a copious Index and Glossary of Musical Terms.
— **Lectures on Musical Analysis.** Embracing Sonata Form, Fugue, etc., Illustrated by the Works of the Classical Masters. 2*nd edition, revised.* Crown 8vo, 7*s.* 6*d.*
— **Musical Art and Study** : Papers for Musicians. Fcap. 8vo, 2*s.*
CHATER (THOMAS). Scientific Voice, Artistic Singing, and Effective Speaking. A Treatise on the Organs of the Voice, their Natural Functions, Scientific Development, Proper Training, and Artistic Use. By THOMAS CHATER. With Diagrams. Wide fcap. 2*s.* 6*d.*
HUNT (H. G. BONAVIA). A Concise History of Music, from the Commencement of the Christian era to the present time. For the use of Students. By REV. H. G. BONAVIA HUNT, Mus. Doc. Dublin ; Warden of Trinity College, London ; and Lecturer on Musical History in the same College. 12*th edition, revised to date* (1893). Fcap. 8vo, 3*s.* 6*d.*

ART.

BARTER (S.) Manual Instruction—Woodwork. By S. BARTER Organizer and Instructor for the London School Board, and to the Joint Committee on Manual Training of the School Board for London, the City and Guilds of London Institute, and the Worshipful Company of Drapers. With over 300 Illustrations. Fcap. 4to, cloth. 7*s.* 6*d.*
BELL (SIR CHARLES). The Anatomy and Philosophy of Expression, as connected with the Fine Arts. By SIR CHARLES BELL, K.H. 7*th edition, revised.* 5*s.*
BRYAN'S Biographical and Critical Dictionary of Painters and Engravers. With a List of Ciphers, Monograms, and Marks. A new Edition, thoroughly Revised and Enlarged. By R. E. GRAVES and WALTER ARMSTRONG. 2 volumes. Imp. 8vo, buckram, 3*l.* 3*s.*
CHEVREUL on Colour. Containing the Principles of Harmony and Contrast of Colours, and their Application to the Arts. 3*rd edition*, with Introduction. Index and several Plates. 5*s.*—With an additional series of 16 Plates in Colours, 7*s.* 6*d.*
DELAMOTTE (P. H.). The Art of Sketching from Nature. By P. H. DELAMOTTE, Professor of Drawing at King's College, London. Illustrated by Twenty-four Woodcuts and Twenty Coloured Plates, arranged progressively, from Water-colour Drawings by PROUT, E. W. COOKE, R.A., GIRTIN, VARLEY, DE WINT, and the Author. *New edition.* Imp. 4to, 21*s.*
FLAXMAN'S CLASSICAL COMPOSITIONS, reprinted in a cheap form for the use of Art Students. Oblong paper covers, 2*s.* 6*d.* each.
The Iliad of Homer. 39 Designs.
The Odyssey of Homer. 34 Designs.
The Tragedies of Æschylus. 36 Designs.
The "Works and Days" and "Theogony" of Hesiod. 37 Designs.
Select Compositions from Dante's Divine Drama. 37 Designs.

Educational Catalogue. 39

FLAXMAN'S Lectures on Sculpture, as delivered before the President and Members of the Royal Academy. With Portrait and 53 plates. 6s.
HEATON (MRS.). A Concise History of Painting. By the late MRS. CHARLES HEATON. *New edition.* Revised by COSMO MONKHOUSE. 5s.
LELAND (C. G.). Drawing and Designing. In a series of Lessons for School use and Self Instruction. By CHARLES G. LELAND, M.A., F.R.L.S. Paper cover, 1s.; or in cloth, 1s. 6d.
— Leather Work: Stamped, Moulded, and Cut, Cuir-Bouillé, Sewn, etc. With numerous Illustrations. Fcap. 4to, 5s.
— Manual of Wood Carving. By CHARLES G. LELAND, M.A., F.R.L.S. Revised by J. J. HOLTZAPFFEL, A.M. INST.C.E. With numerous Illustrations. Fcap. 4to, 5s.
— Metal Work With numerous Illustrations. Fcap. 4to, 5s.
LEONARDO DA VINCI'S Treatise on Painting. Translated from the Italian by J. F. RIGAUD, R.A. With a Life of Leonardo and an Account of his Works, by J. W. BROWN. With numerous Plates. 5s.
MOODY (F. W.). Lectures and Lessons on Art. By the late F. W. MOODY, Instructor in Decorative Art at South Kensington Museum. With Diagrams to illustrate Composition and other matters. *A new and cheaper edition.* Demy 8vo, sewed, 4s. 6d.
WHITE (GLEESON). Practical Designing: A Handbook on the Preparation of Working Drawings, showing the Technical Methods employed in preparing them for the Manufacturer and the Limits imposed on the Design by the Mechanism of Reproduction and the Materials employed. Edited by GLEESON WHITE. Freely Illustrated. *2nd edition.* Crown 8vo, 6s. net.
Contents:—Bookbinding, by H. ORRINSMITH—Carpets, by ALEXANDER MILLAR—Drawing for Reproduction, by the Editor—Pottery, by W. P. RIX—Metal Work, by R. LL. RATHBONE—Stained Glass, by SELWYN IMAGE—Tiles, by OWEN CARTER—Woven Fabrics, Printed Fabrics, and Floorcloths, by ARTHUR SILVER—Wall Papers, by G. C. HAITÉ.

MENTAL, MORAL, AND SOCIAL SCIENCES.

PSYCHOLOGY AND ETHICS.

ANTONINUS (M. Aurelius). The Thoughts of. Translated literally, with Notes, Biographical Sketch, Introductory Essay on the Philosophy, and Index, by GEORGE LONG, M.A. *Revised edition.* Small post 8vo, 3s. 6d., or *new edition on Handmade paper, buckram,* 6s.
BACON'S Novum Organum and Advancement of Learning. Edited, with Notes, by J. DEVEY, M.A. Small post 8vo, 5s.
EPICTETUS. The Discourses of. With the Encheiridion and Fragments. Translated with Notes, a Life of Epictetus, a View of his Philosophy, and Index, by GEORGE LONG, M.A. Small post 8vo, 5s., or *new edition on Handmade paper,* 2 vols., *buckram,* 10s. 6d.
KANT'S Critique of Pure Reason. Translated by J. M. D. MEIKLEJOHN, Professor of Education at St. Andrew's University. Small post 8vo, 5s.
— Prolegomena and Metaphysical Foundations of Science. With Life. Translated by E. BELFORT BAX. Small post 8vo, 5s.

LOCKE'S Philosophical Works. Edited by J. A. ST. JOHN. 2 vols. Small post 8vo, 3s. 6d. each.

RYLAND (F.). The Student's Manual of Psychology and Ethics, designed chiefly for the London B.A. and B.Sc. By F. RYLAND, M.A., late Scholar of St. John's College, Cambridge. Cloth, red edges. 5th edition, revised and enlarged. With lists of books for Students, and Examination Papers set at London University. Crown 8vo, 3s. 6d.

— Ethics: An Introductory Manual for the use of University Students. With an Appendix containing List of Books recommended, and Examination Questions. Crown 8vo, 3s. 6d.

SCHOPENHAUER on the Fourfold Root of the Principle of Sufficient Reason, and On the Will in Nature. Translated by MADAME HILLEBRAND. Small post 8vo, 5s.

— Essays. Selected and Translated. With a Biographical Introduction and Sketch of his Philosophy, by E. BELFORT BAX. Small post 8vo, 5s.

SMITH (Adam). Theory of Moral Sentiments. With Memoir of the Author by DUGALD STEWART. Small post 8vo, 3s. 6d.

SPINOZA'S Chief Works. Translated with Introduction, by R. H. M. ELWES. 2 vols. Small post 8vo, 5s. each.
    Vol. I.—Tractatus Theologico-Politicus—Political Treatise.
       II.—Improvement of the Understanding—Ethics—Letters.

HISTORY OF PHILOSOPHY.

BAX (E. B.). Handbook of the History of Philosophy. By E. BELFORT BAX. 2nd edition, revised. Small post 8vo, 5s.

DRAPER (J. W.). A History of the Intellectual Development of Europe. By JOHN WILLIAM DRAPER, M.D., LL.D. With Index. 2 vols. Small post 8vo, 5s. each

HEGEL'S Lectures on the Philosophy of History. Translated by J. SIBREE, M.A. Small post 8vo, 5s.

LAW AND POLITICAL ECONOMY.

KENT'S Commentary on International Law. Edited by J. T. ABDY, LL.D., Judge of County Courts and Law Professor at Gresham College, late Regius Professor of Laws in the University of Cambridge. 2nd edition, revised and brought down to a recent date. Crown 8vo, 10s. 6d.

LAWRENCE (T. J.). Essays on some Disputed Questions in Modern International Law. By T. J. LAWRENCE, M.A., LL.M. 2nd edition, revised and enlarged. Crown 8vo, 6s.

— Handbook of Public International Law. 2nd edition. Fcap. 8vo, 3s.

MONTESQUIEU'S Spirit of Laws. A New Edition, revised and corrected, with D'Alembert's Analysis, Additional Notes, and a Memoir, by J. V. PRITCHARD, A.M. 2 vols. Small post 8vo, 3s. 6d. each.

RICARDO on the Principles of Political Economy and Taxation. Edited by E. C. K. GONNER, M.A., Lecturer in University College, Liverpool. Small post 8vo, 5s.

SMITH (Adam). The Wealth of Nations. An Inquiry into the Nature and Causes of. Reprinted from the Sixth Edition, with an Introduction by ERNEST BELFORT BAX. 2 vols. Small post 8vo, 3s. 6d. each.

HISTORY.

BOWES (A.). A Practical Synopsis of English History; or, A General Summary of Dates and Events. By ARTHUR BOWES. 10*th edition*. Revised and brought down to the present time. Demy 8vo, 1*s*.

COXE (W.). History of the House of Austria, 1218-1792. By ARCH'DN. COXE, M.A., F.R.S. Together with a Continuation from the Accession of Francis I. to the Revolution of 1848. 4 vols. Small post 8vo. 3*s*. 6*d*. each.

DENTON (W.). England in the Fifteenth Century. By the late REV. W. DENTON, M.A., Worcester College, Oxford. Demy 8vo, 12*s*.

DYER (Dr. T. H.). History of Modern Europe, from the Taking of Constantinople to the Establishment of the German Empire, A.D. 1453-1871. By DR. T. H. DYER. *A new edition*. In 5 vols. £2 12*s*. 6*d*.

GIBBON'S Decline and Fall of the Roman Empire. Complete and Unabridged, with Variorum Notes. Edited by an English Churchman. With 2 Maps. 7 vols. Small post 8vo, 3*s*. 6*d*. each.

GUIZOT'S History of the English Revolution of 1640. Translated by WILLIAM HAZLITT. Small post 8vo, 3*s*. 6*d*.

— History of Civilization, from the Fall of the Roman Empire to the French Revolution. Translated by WILLIAM HAZLITT. 3 vols. Small post 8vo, 3*s*. 6*d*. each.

HENDERSON (E. F.). Select Historical Documents of the Middle Ages. Including the most famous Charters relating to England, the Empire, the Church, etc., from the sixth to the fourteenth centuries. Translated and edited, with Introductions, by ERNEST F. HENDERSON, A.B., A.M., PH.D. Small post 8vo, 5*s*.

— A History of Germany in the Middle Ages. Post 8vo, 7*s*. 6*d*. net.

HOOPER (George). The Campaign of Sedan: The Downfall of the Second Empire, August-September, 1870. By GEORGE HOOPER. With General Map and Six Plans of Battle. Demy 8vo, 14*s*.

— Waterloo: The Downfall of the First Napoleon: a History of the Campaign of 1815. With Maps and Plans. Small post 8vo, 3*s*. 6*d*.

LAMARTINE'S History of the Girondists. Translated by H. T. RYDE. 3 vols. Small post 8vo, 3*s*. 6*d*. each.

— History of the Restoration of Monarchy in France (a Sequel to his History of the Girondists). 4 vols. Small post 8vo, 3*s*. 6*d*. each.

— History of the French Revolution of 1848. Small post 8vo, 3*s*. 6*d*.

LAPPENBERG'S History of England under the Anglo-Saxon Kings. Translated by the late B. THORPE, F.S.A. *New edition*, revised by E. C. OTTÉ. 2 vols. Small post 8vo, 3*s*. 6*d*. each.

LONG (G.). The Decline of the Roman Republic: From the Destruction of Carthage to the Death of Cæsar. By the late GEORGE LONG, M.A. Demy 8vo. In 5 vols. 5*s*. each.

MACHIAVELLI'S History of Florence, and of the Affairs of Italy from the Earliest Times to the Death of Lorenzo the Magnificent : together with the Prince, Savonarola, various Historical Tracts, and a Memoir of Machiavelli. Small post 8vo, 3*s*. 6*d*.

MARTINEAU (H.). History of England from 1800-15. By HARRIET MARTINEAU. Small post 8vo, 3*s*. 6*d*.

— History of the Thirty Years' Peace, 1815-46. 4 vols. Small post 8vo, 3*s*. 6*d*. each.

MAURICE (C. E.). The Revolutionary Movement of 1848-9 in Italy, Austria, Hungary, and Germany. With some Examination of the previous Thirty-three Years. By C. EDMUND MAURICE. With an engraved Frontispiece and other Illustrations. Demy 8vo, 16s.

MENZEL'S History of Germany, from the Earliest Period to 1842. 3 vols. Small post 8vo, 3s. 6d. each.

MICHELET'S History of the French Revolution from its earliest indications to the flight of the King in 1791. Small post 8vo, 3s. 6d.

MIGNET'S History of the French Revolution, from 1789 to 1814. Small post 8vo, 3s. 6d.

PARNELL (A.). The War of the Succession in Spain during the Reign of Queen Anne, 1702-1711. Based on Original Manuscripts and Contemporary Records. By COL. THE HON. ARTHUR PARNELL, R.E. Demy 8vo, 14s. With Map, etc.

RANKE (L.). History of the Latin and Teutonic Nations, 1494-1514. Translated by P. A. ASHWORTH. Small post 8vo, 3s. 6d.

— History of the Popes, their Church and State, and especially of their conflicts with Protestantism in the 16th and 17th centuries. Translated by E. FOSTER. 3 vols. Small post 8vo, 3s. 6d. each.

— History of Servia and the Servian Revolution. Translated by MRS. KERR. Small post 8vo, 3s. 6d.

SIX OLD ENGLISH CHRONICLES: viz., Asser's Life of Alfred and the Chronicles of Ethelwerd, Gildas, Nennius, Geoffrey of Monmouth, and Richard of Cirencester. Edited, with Notes and Index, by J. A. GILES, D.C.L. Small post 8vo, 5s.

STRICKLAND (Agnes). The Lives of the Queens of England; from the Norman Conquest to the Reign of Queen Anne. By AGNES STRICKLAND. 6 vols. 5s. each.

— The Lives of the Queens of England. Abridged edition for the use of Schools and Families, Post 8vo, 6s. 6d.

THIERRY'S History of the Conquest of England by the Normans; its Causes, and its Consequences in England, Scotland, Ireland, and the Continent. Translated from the 7th Paris edition by WILLIAM HAZLITT. 2 vols. Small post 8vo, 3s. 6d. each.

WRIGHT (H. F.). The Intermediate History of England, with Notes, Supplements, Glossary, and a Mnemonic System. For Army and Civil Service Candidates. By H. F. WRIGHT, M.A., LL.M. Crown 8vo, 6s.

For other Works of value to Students of History, see Catalogue of Bohn's Libraries, sent post-free on application.

DIVINITY, ETC.

ALFORD (DEAN). Greek Testament. With a Critically revised Text, a digest of Various Readings, Marginal References to verbal and idiomatic usage, Prolegomena, and a Critical and Exegetical Commentary. For the use of theological students and ministers. By the late HENRY ALFORD, D.D., Dean of Canterbury. 4 vols. 8vo. £5 2s. Sold separately.

— The New Testament for English Readers. Containing the Authorized Version, with additional Corrections of Readings and Renderings, Marginal References, and a Critical and Explanatory Commentary. In 2 vols. £2 14s. 6d. Also sold in 4 parts separately.

AUGUSTINE de Civitate Dei. Books XI. and XII. By the REV. HENRY D. GEE, B.D., F.S.A. I. Text only. 2s. II. Introduction and Translation. 3s.

BARRETT (A. C.). Companion to the Greek Testament. By the late A. C. BARRETT, M.A., Caius College, Cambridge. 5*th edition*. Fcap. 8vo, 5s.

BARRY (BP.). Notes on the Catechism. For the use of Schools. By the RT. REV. BISHOP BARRY, D.D. 10*th edition*. Fcap. 2s.

BLEEK. Introduction to the Old Testament. By FRIEDRICH BLEEK. Edited by JOHANN BLEEK and ADOLF KAMPHAUSEN. Translated from the second edition of the German by G. H. VENABLES, under the supervision of the REV. E. VENABLES, Residentiary Canon of Lincoln. *2nd edition*, with Corrections. With Index. 2 vols. small post 8vo, 5s. each.

BUTLER (BP.). Analogy of Religion. With Analytical Introduction and copious Index, by the late RT. REV. DR. STEERE. Fcap. 3s. 6d.

EUSEBIUS. Ecclesiastical History of Eusebius Pamphilus, Bishop of Cæsarea. Translated from the Greek by REV. C. F. CRUSE, M.A. With Notes, a Life of Eusebius, and Chronological Table. Sm. post 8vo, 5s.

GREGORY (DR.). Letters on the Evidences, Doctrines, and Duties of the Christian Religion. By DR. OLINTHUS GREGORY, F.R.A.S. Small post 8vo, 3s. 6d.

HUMPHRY (W. G.). Book of Common Prayer. An Historical and Explanatory Treatise on the. By W. G. HUMPHRY, B.D., late Fellow of Trinity College, Cambridge, Prebendary of St. Paul's, and Vicar of St. Martin's-in-the-Fields, Westminster. 6*th edition*. Fcap. 8vo, 2s. 6d.
Cheap Edition, for Sunday School Teachers. 1s.

JOSEPHUS (FLAVIUS). The Works of. WHISTON'S Translation. Revised by REV. A. R. SHILLETO, M.A. With Topographical and Geographical Notes by COLONEL SIR C. W. WILSON, K.C.B. 5 vols. 3s. 6d. each.

LUMBY (DR.). The History of the Creeds. I. Ante-Nicene. II. Nicene and Constantinopolitan. III. The Apostolic Creed. IV. The Quicunque, commonly called the Creed of St. Athanasius. By J. RAWSON LUMBY, D.D., Norrisian Professor of Divinity, Fellow of St. Catherine's College, and late Fellow of Magdalene College, Cambridge. 3*rd edition*, revised. Crown 8vo, 7s. 6d.

— Compendium of English Church History, from 1688-1830. With a Preface by J. RAWSON LUMBY, D.D. Crown 8vo, 6s.

MACMICHAEL (J. F.). The New Testament in Greek. With English Notes and Preface, Synopsis, and Chronological Tables. By the late REV. J. F. MACMICHAEL. Fcap. 8vo (730 pp.), 4s. 6d.
Also the Four Gospels, and the Acts of the Apostles, separately. In paper wrappers, 6d. each.

MILLER (E). Guide to the Textual Criticism of the New Testament. By REV. E. MILLER, M.A., Oxon, Rector of Bucknell, Bicester. Crown 8vo, 4s.

NEANDER (DR. A.). History of the Christian Religion and Church. Translated by J. TORREY. 10 vols. small post 8vo, 3s. 6d. each.

— Life of Jesus Christ. Translated by J. MCCLINTOCK and C. BLUMENTHAL. Small post 8vo, 3s. 6d.

— History of the Planting and Training of the Christian Church by the Apostles. Translated by J. E. RYLAND. 2 vols. 3s. 6d. each.

— Lectures on the History of Christian Dogmas. Edited by DR. JACOBI. Translated by J. E. RYLAND. 2 vols. small post 8vo, 3s. 6d. each.

NEANDER (DR. A.). Memorials of Christian Life in the Early and Middle Ages. Translated by J. E. RYLAND. Small post 8vo, 3s. 6d.

PEARSON (BP.). On the Creed. Carefully printed from an Early Edition. Edited by E. WALFORD, M.A. Post 8vo, 5s.

PEROWNE (BP.). The Book of Psalms. A New Translation, with Introductions and Notes, Critical and Explanatory. By the RIGHT REV. J. J. STEWART PEROWNE, D.D., Bishop of Worcester. 8vo. Vol. I. 8th edition, revised. 18s. Vol. II. 7th edition, revised. 16s.

— The Book of Psalms. Abridged Edition for Schools. Crown 8vo, 7th edition. 10s. 6d.

SADLER (M. F.). The Church Teacher's Manual of Christian Instruction. Being the Church Catechism, Expanded and Explained in Question and Answer. For the use of the Clergyman, Parent, and Teacher. By the REV. M. F. SADLER, Prebendary of Wells, and Rector of Honiton. 43rd thousand. 2s. 6d.

*** A Complete List of Prebendary Sadler's Works will be sent on application.

SCRIVENER (DR.). A Plain Introduction to the Criticism of the New Testament. With Forty-four Facsimiles from Ancient Manuscripts. For the use of Biblical Students. By the late F. H. SCRIVENER, M.A., D.C.L., LL.D., Prebendary of Exeter. 4th edition, thoroughly revised, by the REV. E. MILLER, formerly Fellow and Tutor of New College, Oxford. 2 vols. demy 8vo, 32s.

— Novum Testamentum Græce, Textus Stephanici, 1550. Accedunt variae lectiones editionum Bezae, Elzeviri, Lachmanni, Tischendorfii, Tregellesii, curante F. H. A. SCRIVENER, A.M., D.C.L., LL.D. Revised edition. 4s. 6d.

— Novum Testamentum Græce [Editio Major] textus Stephanici, A.D. 1556. Cum variis lectionibus editionum Bezae, Elzeviri, Lachmanni, Tischendorfii, Tregellesii, Westcott-Hortii, versionis Anglicanæ emendatorum curante F. H. A. SCRIVENER, A.M., D.C.L., LL.D., accedunt parallela s. scripturæ loca. Small post 8vo. 2nd edition. 7s. 6d.

An Edition on writing-paper, with margin for notes. 4to, half bound, 12s.

WHEATLEY. A Rational Illustration of the Book of Common Prayer. Being the Substance of everything Liturgical in Bishop Sparrow, Mr. L'Estrange, Dr. Comber, Dr. Nicholls, and all former Ritualist Commentators upon the same subject. Small post 8vo, 3s. 6d.

WHITAKER (C.). Rufinus and His Times. With the Text of his Commentary on the Apostles' Creed and a Translation. To which is added a Condensed History of the Creeds and Councils. By the REV. CHARLES WHITAKER, B.A., Vicar of Natland, Kendal. Demy 8vo, 5s.

Or in separate Parts.—1. Latin Text, with Various Readings, 2s. 6d. 2. Summary of the History of the Creeds, 1s. 6d. 3. Charts of the Heresies of the Times preceding Rufinus, and the First Four General Councils, 6d. each.

— St. Augustine: De Fide et Symbolo—Sermo ad Catechumenos. St. Leo ad Flavianum Epistola—Latin Text, with Literal Translation, Notes, and History of Creeds and Councils. 5s. Also separately, Literal Translation. 2s.

— Student's Help to the Prayer-Book. 3s.

Educational Catalogue. 45

SUMMARY OF SERIES.

	PAGE
BIBLIOTHECA CLASSICA	45
PUBLIC SCHOOL SERIES	45
CAMBRIDGE GREEK AND LATIN TEXTS	46
CAMBRIDGE TEXTS WITH NOTES	46
GRAMMAR SCHOOL CLASSICS	46
PRIMARY CLASSICS	47
BELL'S CLASSICAL TRANSLATIONS	47
CAMBRIDGE MATHEMATICAL SERIES	47
CAMBRIDGE SCHOOL AND COLLEGE TEXT BOOKS	49
FOREIGN CLASSICS	49
MODERN FRENCH AUTHORS	49
MODERN GERMAN AUTHORS	49
GOMBERT'S FRENCH DRAMA	50
BELL'S MODERN TRANSLATIONS	50
BELL'S ENGLISH CLASSICS	50
HANDBOOKS OF ENGLISH LITERATURE	50
TECHNOLOGICAL HANDBOOKS	50
BELL'S AGRICULTURAL SERIES	50
BELL'S READING BOOKS AND GEOGRAPHICAL READERS	50

BIBLIOTHECA CLASSICA.

AESCHYLUS. By DR. PALEY. 8s.
CICERO. By G. LONG. Vols. I. and II. 8s. each.
DEMOSTHENES. By R. WHISTON. 2 Vols. 8s. each.
EURIPIDES. By DR PALEY. Vols. II. and III. 8s. each.
HERODOTUS. By DR. BLAKESLEY. 2 Vols. 12s.
HESIOD. By DR. PALEY. 5s.
HOMER. By DR. PALEY. 2 Vols. 14s.
HORACE. By A. G. MACLEANE. 8s.
PLATO. Phaedrus. By DR. THOMPSON. 5s.
SOPHOCLES. Vol. I. By F. H. BLAYDES. 5s.
— Vol. II. By DR. PALEY. 6s.
VIRGIL. By CONINGTON AND NETTLESHIP. 3 Vols. 10s. 6d. each.

PUBLIC SCHOOL SERIES.

ARISTOPHANES. Peace. By DR. PALEY. 4s. 6d.
— Acharnians. By DR. PALEY. 4s. 6d.
— Frogs. By DR. PALEY. 4s. 6d.
CICERO. Letters to Atticus. Book I. By A. PRETOR. 4s. 6d.
DEMOSTHENES. De Falsa Legatione. By R. SHILLETO. 6s.
— Adv. Leptinem. By B. W. BEATSON. 3s. 6d.
LIVY. Books XXI. and XXII. By L. D. DOWDALL. 3s. 6d. each.
PLATO. Apology of Socrates and Crito. By DR. W. WAGNER. 3s. 6d. and 2s. 6d.
— Phaedo. By DR. W. WAGNER. 5s. 6d.
— Protagoras. By W. WAYTE. 4s. 6d.
— Gorgias. By DR. THOMPSON. 6s.
— Euthyphro. By G. H. WELLS. 3s.
— Euthydemus. By G. H. WELLS. 4s.
— Republic. By G. H. WELLS. 5s.
PLAUTUS. Aulularia. By DR. W. WAGNER. 4s. 6d.
— Trinummus. By DR. W. WAGNER. 4s. 6d.
— Menaechmei. By DR. W. WAGNER. 4s. 6d.
— Mostellaria. By E. A. SONNENSCHEIN. 5s.

PUBLIC SCHOOL SERIES—continued.

SOPHOCLES. Trachiniae. By A. PRETOR. 4s. 6d.
— Oedipus Tyrannus. By B. H. KENNEDY. 5s.
TERENCE. By DR. W. WAGNER. 7s. 6d.
THEOCRITUS. By DR. PALEY. 4s. 6d.
THUCYDIDES. Book VI. By T. W. DOUGAN. 3s. 6d.

CAMBRIDGE GREEK AND LATIN TEXTS.

AESCHYLUS. By DR. PALEY. 2s.
CAESAR. By G. LONG. 1s. 6d.
CICERO. De Senectute, de Amicitia, et Epistolae Selectae. By G. LONG. 1s. 6d.
— Orationes in Verrem. By G. LONG. 2s. 6d.
EURIPIDES. By DR. PALEY. 3 Vols. 2s. each.
HERODOTUS. By DR. BLAKESLEY. 2 Vols. 2s. 6d. each.
HOMER'S Iliad. By DR. PALEY. 1s. 6d.
HORACE. By A. J. MACLEANE. 1s. 6d.
JUVENAL AND PERSIUS. By A. J. MACLEANE. 1s. 6d.
LUCRETIUS. By H. A. J. MUNRO. 2s.
SOPHOCLES. By DR. PALEY. 2s. 6d.
TERENCE. By DR. W. WAGNER. 2s.
THUCYDIDES. By DR. DONALDSON. 2 Vols. 2s. each.
VIRGIL. By PROF. CONINGTON. 2s.
XENOPHON. By J. F. MACMICHAEL. 1s. 6d.
NOVUM TESTAMENTUM GRAECE. By DR. SCRIVENER. 4s. 6d.

CAMBRIDGE TEXTS WITH NOTES.

AESCHYLUS. By DR. PALEY. 6 Vols. 1s. 6d. each.
EURIPIDES. By DR. PALEY. 13 Vols. (Ion, 2s.) 1s. 6d. each.
HOMER'S Iliad. By DR. PALEY. 1s.
SOPHOCLES. By DR. PALEY. 5 Vols. 1s. 6d. each.
XENOPHON. Hellenica. By REV. L. D. DOWDALL. Books I. and II. 2s. each.
— Anabasis. By J. F. MACMICHAEL. 6 Vols. 1s. 6d. each.
CICERO. De Senectute, de Amicitia, et Epistolae Selectae. By G. LONG. 3 Vols. 1s. 6d. each.
OVID. Selections. By A. J. MACLEANE. 1s. 6d.
— Fasti. By DR. PALEY. 3 Vols. 2s. each.
TERENCE. By DR. W. WAGNER. 4 Vols. 1s. 6d. each.
VIRGIL. By PROF. CONINGTON. 12 Vols. 1s. 6d. each.

GRAMMAR SCHOOL CLASSICS.

CAESAR, De Bello Gallico. By G. LONG. 4s., or in 3 parts, 1s. 6d. each.
CATULLUS, TIBULLUS, and PROPERTIUS. By A. H. WRATISLAW, and F. N. SUTTON. 2s. 6d.
CORNELIUS NEPOS. By J. F. MACMICHAEL. 2s.
CICERO. De Senectute, De Amicitia, and Select Epistles. By G. LONG. 3s.
HOMER. Iliad. By DR. PALEY. Books I.-XII. 4s. 6d., or in 2 Parts, 2s. 6d. each.
HORACE. By A. J. MACLEANE. 3s. 6d., or in 2 Parts, 2s. each.
JUVENAL. By HERMAN PRIOR. 3s. 6d.
MARTIAL. By DR. PALEY and W. H. STONE. 4s. 6d.
OVID. Fasti. By DR. PALEY. 3s. 6d., or in 3 Parts, 1s. 6d. each.
SALLUST. Catilina and Jugurtha. By G. LONG and J. G. FRAZER. 3s. 6d., or in 2 Parts, 2s. each.
TACITUS. Germania and Agricola. By P. FROST. 2s. 6d.
VIRGIL. CONINGTON's edition abridged. 2 Vols. 4s. 6d. each, or in 9 Parts, 1s. 6d. each.
— Bucolics and Georgics. CONINGTON's edition abridged. 3s.
XENOPHON. By J. F. MACMICHAEL. 3s. 6d., or in 4 Parts, 1s. 6d. each.
— Cyropaedia. By G. M. GORHAM. 3s. 6d., or in 2 Parts, 1s. 6d. each.
— Memorabilia. By PERCIVAL FROST. 3s.

PRIMARY CLASSICS.

EASY SELECTIONS FROM CAESAR. By A. M. M. STEDMAN. 1s.
EASY SELECTIONS FROM LIVY. By A. M. M. STEDMAN. 1s. 6d.
EASY SELECTIONS FROM HERODOTUS. By A. G. LIDDELL. 1s. 6d.

BELL'S CLASSICAL TRANSLATIONS.

AESCHYLUS. By WALTER HEADLAM. 6 Vols. [*In the press.*
ARISTOPHANES. Acharnians. By W. H. COVINGTON. 1s.
CAESAR'S Gallic War. By W. A. MCDEVITTE. 2 Vols. 1s. each.
CICERO. Friendship and Old Age. By G. H. WELLS. 1s.
DEMOSTHENES. On the Crown. By C. RANN KENNEDY. 1s.
EURIPIDES. 14 Vols. By E. P. COLERIDGE. 1s. each.
LIVY. Books I.-IV. By J. H. FREESE. 1s. each.
— Book V. By E. S. WEYMOUTH. 1s.
— Book IX. By F. STORR. 1s.
LUCAN: The Pharsalia. Book I. By F. CONWAY. 1s.
SOPHOCLES. 7 Vols. By E. P. COLERIDGE. 1s. each.
VIRGIL. 6 Vols. By A. HAMILTON BRYCE. 1s. each.

CAMBRIDGE MATHEMATICAL SERIES.

ARITHMETIC. By C. PENDLEBURY. 4s. 6d., or in 2 Parts, 2s. 6d. each. Key to Part II. 7s. 6d. net.
EXAMPLES IN ARITHMETIC. By C. PENDLEBURY. 3s., or in 2 Parts, 1s. 6d. and 2s.
ARITHMETIC FOR INDIAN SCHOOLS. By PENDLEBURY and TAIT. 3s.
ELEMENTARY ALGEBRA. By J. T. HATHORNTHWAITE. 2s.
CHOICE AND CHANCE. By W. A. WHITWORTH. 6s.
EUCLID. By H. DEIGHTON. 4s. 6d., or Book I., 1s.; Books I. and II., 1s. 6d.; Books I.-III., 2s. 6d.; Books III. and IV., 1s. 6d.
— KEY. 5s. net.
EXERCISES ON EUCLID, &c. By J. MCDOWELL. 6s.
ELEMENTARY TRIGONOMETRY. By DYER and WHITCOMBE. 4s. 6d.
PLANE TRIGONOMETRY. By T. G. VYVYAN. 3s. 6d.
ANALYTICAL GEOMETRY FOR BEGINNERS Part I. By T. G. VYVYAN. 2s. 6d.
ELEMENTARY GEOMETRY OF CONICS. By DR. TAYLOR. 4s. 6d.
GEOMETRICAL CONIC SECTIONS. By H. G. WILLIS. 5s.
SOLID GEOMETRY. By W. S. ALDIS. 6s.
GEOMETRICAL OPTICS. By W. S. ALDIS. 4s.
ROULETTES AND GLISSETTES. By DR. W. H. BESANT. 5s.
ELEMENTARY HYDROSTATICS. By DR. W. H. BESANT. 4s. 6d. Solutions. 5s.
HYDROMECHANICS. Part I. Hydrostatics. By DR. W. H. BESANT. 5s.
DYNAMICS. By DR. W. H. BESANT. 10s. 6d.
RIGID DYNAMICS. By W. S. ALDIS. 4s.
ELEMENTARY DYNAMICS. By DR. W. GARNETT. 6s.
ELEMENTARY TREATISE ON HEAT. By DR. W. GARNETT. 4s. 6d.
ELEMENTS OF APPLIED MATHEMATICS. By C. M. JESSOP. 6s.
PROBLEMS IN ELEMENTARY MECHANICS. By W. WALTON. 6s.
EXAMPLES IN ELEMENTARY PHYSICS. By W. GALLATLY. 4s.
MATHEMATICAL EXAMPLES. By DYER and PROWDE SMITH. 6s.

CAMBRIDGE SCHOOL AND COLLEGE TEXT BOOKS.

ARITHMETIC. By C. ELSEE. 3s. 6d.
By A. WRIGLEY. 3s. 6d.
EXAMPLES IN ARITHMETIC. By WATSON and GOUDIE. 2s. 6d.
ALGEBRA By C. ELSEE. 4s.
EXAMPLES IN ALGEBRA. By MACMICHAEL and PROWDE SMITH. 3s. 6d. and 4s. 6d.
PLANE ASTRONOMY. By P. T. MAIN. 4s.
GEOMETRICAL CONIC SECTIONS. By DR. W. H. BESANT. 4s. 6d.
STATICS. By BISHOP GOODWIN. 3s.
NEWTON'S Principia. By EVANS and MAIN. 4s.
ANALYTICAL GEOMETRY. By T. G. VYVYAN. 4s. 6d.
COMPANION TO THE GREEK TESTAMENT. By A. C. BARRETT. 5s.
TREATISE ON THE BOOK OF COMMON PRAYER. By W. G. HUMPHRY. 2s. 6d.
TEXT BOOK OF MUSIC. By H. C. BANISTER. 5s.
CONCISE HISTORY OF MUSIC. By DR. H. G. BONAVIA HUNT. 3s. 6d.

FOREIGN CLASSICS.

FÉNELON'S Télémaque. By C. J. DELILLE. 2s. 6d.
LA FONTAINE'S Select Fables. By F. E. A. GASC. 1s. 6d.
LAMARTINE'S Le Tailleur de Pierres de Saint-Point. By J. BOÏELLE. 1s. 6d.
SAINTINE'S Picciola. By DR. DUBEC. 1s. 6d.
VOLTAIRE'S Charles XII. By L. DIRY. 1s. 6d.
GERMAN BALLADS. By C. L. BIELEFELD. 1s. 6d.
GOETHE'S Hermann und Dorothea. By E. BELL and E. WÖLFEL. 1s. 6d.
SCHILLER'S Wallenstein. By DR. BUCHHEIM. 5s., or in 2 Parts, 2s. 6d. each.
— Maid of Orleans. By DR. W. WAGNER. 1s. 6d.
— Maria Stuart. By V. KASTNER. 1s. 6d.

MODERN FRENCH AUTHORS.

BALZAC'S Ursule Mirouët. By J. BOÏELLE. 3s.
CLARÉTIE'S Pierrille. By J. BOÏELLE. 2s. 6d.
DAUDET'S La Belle Nivernaise. By J. BOÏELLE. 2s.
GREVILLE'S Le Moulin Frappier. By J. BOÏELLE. 3s.
HUGO'S Bug Jargal. By J. BOÏELLE. 3s.

MODERN GERMAN AUTHORS.

HEY'S Fabeln fur Kinder. By PROF. LANGE. 1s. 6d.
— — with Phonetic Transcription of Text, &c. 2s.
FREYTAG'S Soll und Haben. By W. H. CRUMP. 2s. 6d.
BENEDIX'S Doktor Wespe. By PROF. LANGE. 2s. 6d.
HOFFMANN'S Meister Martin. By PROF. LANGE. 1s. 6d.
HEYSE'S Hans Lange. By A. A. MACDONELL. 2s.
AUERBACH'S Auf Wache, and Roquette's Der Gefrorene Kuss. By A. A. MACDONELL. 2s.
MOSER'S Der Bibliothekar. By PROF. LANGE. 2s.
EBERS' Eine Frage. By F. STORR. 2s.
FREYTAG'S Die Journalisten. By PROF. LANGE. 2s. 6d.
GUTZKOW'S Zopf und Schwert. By PROF. LANGE. 2s. 6d.
GERMAN EPIC TALES. By DR. KARL NEUHAUS. 2s. 6d.
SCHEFFEL'S Ekkehard. By DR. H. HAGER. 3s.

The following Series are given in full in the body of the Catalogue.

GOMBERT'S French Drama. *See page* 31.
BELL'S Modern Translations. *See page* 34.
BELL'S English Classics. *See pp.* 24, 25.
HANDBOOKS OF ENGLISH LITERATURE. *See page* 26.
TECHNOLOGICAL HANDBOOKS *See page* 37.
BELL'S Agricultural Series. *See page* 36.
BELL'S Reading Books and Geographical Readers. *See pp.* 25, 26.

www.ingramcontent.com/pod-product-compliance
Lightning Source LLC
Chambersburg PA
CBHW031731230426
43669CB00007B/320